ENGINEERING INTERVENTIONS IN FOODS AND PLANTS

Innovations in Agricultural and Biological Engineering

ENGINEERING INTERVENTIONS IN FOODS AND PLANTS

Edited by

Deepak Kumar Verma, PhD
Megh R. Goyal, PhD

Apple Academic Press Inc.
3333 Mistwell Crescent
Oakville, ON L6L 0A2 Canada

Apple Academic Press Inc.
9 Spinnaker Way
Waretown, NJ 08758 USA

© 2018 by Apple Academic Press, Inc.

First issued in paperback 2021

Exclusive worldwide distribution by CRC Press, a member of Taylor & Francis Group
No claim to original U.S. Government works

ISBN 13: 978-1-77-463641-1 (pbk)
ISBN 13: 978-1-77-188596-6 (hbk)

Library and Archives Canada Cataloguing in Publication

Engineering interventions in foods and plants / edited by Deepak Kumar Verma, PhD, Megh R. Goyal, PhD.
(Innovations in agricultural and biological engineering)
Includes bibliographical references and index.
Issued in print and electronic formats.
ISBN 978-1-77188-596-6 (hardcover).--ISBN 978-1-315-19467-7 (PDF)
1. Food industry and trade. 2. Agricultural engineering. 3. Bioengineering. 4. Food crops. 5. Plants, Edible. I. Goyal, Megh Raj, editor II. Verma, Deepak Kumar, 1986-, editor III. Series: Innovations in agricultural and biological engineering
TP370.E54 2017   664   C2017-905007-9   C2017-905008-7

Library of Congress Cataloging-in-Publication Data

Names: Verma, Deepak Kumar, 1986- editor. | Goyal, Megh Raj, editor.
Title: Engineering interventions in foods and plants / editors: Deepak Kumar Verma, PhD; Megh R. Goyal, PhD.
Description: Waretown, NJ : Apple Academic Press, 2017. | Includes bibliographical references and index.
Identifiers: LCCN 2017033109 (print) | LCCN 2017034333 (ebook) | ISBN 9781315194677 (ebook) | ISBN 9781771885966 (hardcover : alk. paper)
Subjects: LCSH: Agricultural processing. | Food--Technological innovations. | Plant products--Technological innovations.
Classification: LCC S698 (ebook) | LCC S698 .E55 2017 (print) | DDC 630.72/4--dc23
LC record available at https://lccn.loc.gov/2017033109

Apple Academic Press also publishes its books in a variety of electronic formats. Some content that appears in print may not be available in electronic format. For information about Apple Academic Press products, visit our website at **www.appleacademicpress.com** and the CRC Press website at **www.crcpress.com**

CONTENTS

LIST OF CONTRIBUTORS

Asaad Rehman Saeed Al-Hilphy
Assistant Professor, Department of Food Science, College of Agriculture, University of Basrah, Basrah City, Iraq; Mobile: +00-96-47702696458; E-mail: aalhilphy@yahoo.co.uk

Haider I. Ali
Assistant Professor, Department of Food Science, College of Agriculture, University of Basrah, Basra City, Iraq; Mobile: +00-96-47703131398; E-mail: haiderr_2004@yahoo.com

Sudhanshi Billoria
PhD Research Scholar, Department of Agricultural and Food Engineering, Indian Institute of Technology, Kharagpur – 721302, West Bengal, India; Mobile: +91-8768126479; E-mail: sudharihant@gmail.com

Deepali Chauhan
Scientist (Home science), KVK (CSAUA&T, Kanpur), Dariyapur, Raibareli – 229001, Uttar Pradesh, India; Tel.: +91-5352001732; Mobile: +91-9839946033; E-mail: deepali_chauhan20@rediffmail.com

Annu Goel
Research Associate, National Ganga River Basin Authority, Central Pollution Control Board, Parivesh Bhawan, East Arjun Nagar, Delhi – 110032; Mobile: +91-9711104743; E-mail: anugoel.micro@gmail.com

Megh R. Goyal
Retired Professor in Agricultural and Biomedical Engineering, University of Puerto Rico – Mayaguez Campus; and Senior Technical Editor-in-Chief in Agriculture Sciences and Biomedical Engineering, Apple Academic Press Inc., Mailing Address: PO Box 86, Rincon – PR – 00677, USA; E-mail: goyalmegh@gmail.com

Arpana H. Jobanputra
Assistant Professor, Department of Microbiology, PSGVPM's, SIP Arts, GBP Science & STSKVS Commerce College, Shahada – 425409, Maharashtra, India; Mobile: +91-9822401695; Tel.: +91-2565229576; Fax: +912565229576; E-mail: arpana_j12@rediffmail.com

Dipendra Kumar Mahato
Senior Research Fellow, Indian Agricultural Research Institute (IARI), Pusa Campus, New Delhi – 110012, India; Mobile: +91-9911891494 or 9958921936; E-mail: kumar.dipendra2@gmail.com

Ghassan. F. Mohsin
Vocational Teacher, Maysan City, Iraq; Mobile: +00-4915214130015; E-mail: ghassanmohsin@ymail.com

Ashwini G. Patil
Assistant Professor (Microbiology), Faculty of Science, R.C. Patel A.C.S. College, Shirpur – 425405, Maharashtra, India; E-mail: ashwinipatil447@gmail.com

Pravin O. Patil
Assistant Professor, Department of Pharmaceutical Chemistry, H. R. Patel Institute of Pharmaceutical Education and Research, Shirpur – 425405 – Maharashtra, India; E-mail: popatil@hrpatelpharmacy.co.in

Kushal Raj
Assistant Scientist, Department of Plant Pathology, Chaudhary Charan Singh Haryana Agricultural University, Hisar – 125004, Haryana, India; Mobile: +91-9416263020; E-mail: kushalraj2008@gmail.com

Vishal Singh
Assistant Professor & Head, Department of Agricultural Engineering, M. S. Swaminathan School of Agriculture, Centurion University of Technology and Management, Jatni P.O., District Khurda – 752050, Odisha, India; Mobile: +00-91-8093872582, +00-91-8348521736; E-mail: vishalsinghiitkgp87@gmail.com, vishalsinghiitkgp@cutm.ac.in

Prem Prakash Srivastav
Associate Professor, Agricultural and Food Engineering Department, Indian Institute of Technology, Kharagpur – 721302, West Bengal, India; Tel.: +91-3222281673; E-mail: pps@agfe.iitkgp.ernet.in

Ajay Kumar Swarnakar
PhD Research Scholar, Agricultural and Food Engineering Department, Indian Institute of Technology, Kharagpur – 721302, West Bengal, India; Mobile: +91-8101766639; E-mail: aksw11@gmail.com

Deepak Kumar Verma
PhD Research Scholar, Agricultural and Food Engineering Department, Indian Institute of Technology, Kharagpur – 721302, West Bengal, India; Tel.: +91-3222281673; Mobile: +91--7407170260; Fax: +91-3222282224; E-mail: deepak.verma@agfe.iitkgp.ernet.in; rajadkv@rediffmail.com

Leela Wati
Principal Scientist, Department of Microbiology, Chaudhary Charan Singh Haryana Agricultural University, Hisar – 125004, Haryana, India; Mobile: +91-9416397853; Tel.: +91-1662-285292; E-mail: lwkaj@gmail.com

LIST OF ABBREVIATIONS

AA	amino acid
AAO	ascorbic acid oxidase
ADP	adenosine diphosphate
AFEX	ammonia fiber explosion
Akt	proteinkinase B
ATP	adenosine triphosphate
a_w	water activity
BOD	biochemical oxygen demand
$(C_6H_{10}O_5)_n$	cellulose
$C_{12}H_{17}N_4OS^+$	thiamine
$C_{12}H_{22}O_{11}$	sucrose
$C_{17}H_{20}N_4O_6$	riboflavin
$C_{19}H_{19}N_7O_6$	folic acid or folate or folacin
C_2H_4	ethene
C_2H_5OH	ethanol
$C_{63}H_{88}CoN_{14}O_{14}P$	cobalamin
$C_6H_{12}O_6$	fructose
$C_6H_{12}O_6$	glucose
$C_6H_8O_6$	ascorbate
$C_6NH_5O_2$	niacin or nicotinic acid
$C_8H_{11}NO_3$	pyridoxine or pyridoxol
$C_9H_{17}NO_5$	panthothenic acid or pantothenate
Ca	calcium
CAPS	cover and plinth storage structure
CAT	catalase
CBP	consolidated bioprocessing
CDS	condensed distillers solubles
CG	catechin gallate
CH_3CH_2OH	ethanol
CNS	central nervous system
CO_2	carbon dioxide

COD	chemical oxygen demand
CVD	cardiovascular diseases
DDGS	dried distillers grains with solubles
EC	epicatechin
ECG	epicatechin-3-gallate
ED	pathway enter-doudoroff pathway
EGC	epigallocatechin
EGCG	epigallocatechin gallate
EMP	embden-meyerhof pathway
FAO	Food and Agriculture Organization
FCC	food chemical codex
FCI	Food Corporation of India
Fe	iron
FID	flame ionization detector
GC	gallocatechin
GC	gas chromatography
GC-MS	gas chromatography mass spectrometry
GCG	gallocatechin gallate
GCO	gas chromatography olfactometry
GLFs	green leafy vegetables
GNS	grain neutral spirits
GRAS	generally recognized as safe
H_2O_2	hydrogen peroxide
H_2SO_4	sulfuric acid
HACCP	hazard analysis critical control point
HCl	hydrochloric acid
HD	hydro (water) distillation
HIV	human immunodeficiency virus
HPLC	high performance liquid chromatography
HSH	hydrogenated starch hydrolysate
HTST	high-temperature short-time
IR	infrared spectroscopy
ITCs	isothiocyanates
ITU	International Telecommunication Union
IUPAC	International Union of Pure and Applied Chemistry
K	potassium

LDPE	low-density polyethylene
LTLT	low-temperature long-time
MAWD	microwave-assisted water distillation
MC	moisture content
Mg	magnesium
$MgCO_3$	magnesium carbonate
Mn	manganese
MS	multiple sclerosis
MS	mass spectrometry
MSSP	minimum safe sterilization process
MSW	municipal solid wastes
Na	sodium
NaOH	sodium hydroxide
NDEA	nitrosodiethylamine
OH	ohmic heating
OHHD	ohmic heating hydro (water) distillation
P	phosphorus
PID	photo ionization detector
PO	peroxidase
ppm	parts per million
PPO	polyphenol oxidase
PVP	polyvinyl pyrrolidone
RH	relative humidity
SCMC	sodium carboxy methyl cellulose
SHF	separate hydrolysis and fermentation
SPC	system performance coefficient
SSCF	simultaneous saccharification and co-fermentation
SSF	simultaneous saccharification and fermentation
TWD	traditional water distillation
UHT	ultrahigh temperature
USA	United State of America
USDA	United States Department of Agriculture
UVR	ultra-violet radiation
WHO	World Health Organization
Zn	zinc

PREFACE 1 BY DEEPAK KUMAR VERMA

Food is an integral component of life and meets physiological and nutritional demands for human existence. Foods are accepted by living beings depending on their quality, quantity and variety. The main source of food is agriculture, especially crop plants, which entrap solar energy and converts it into chemical energy in the form of biomass and is further used for food, feed, fiber and fuel. However, according to the present world scenario, about 30–50% of the total food produced gets spoiled or squandered before consumption, due to lack of proper processing and storage technology. This causes lack of food and a hike of its price, resulting in starvation in developed countries. Some other issues are drastic changes and reduction in biodiversity, over-fertilization causing aquatic eutrophication by nitrogenous and phosphorous substances, enteric fermentation and use of fossil fuels resulting to global warming, water shortages due to irrigation, eco-toxicity, side-effects of pesticides, etc. Thus, destruction of food and lack of storage and supply has resulted in global malnutrition and protein, calorie and mineral deficiencies in 800 million people, causing risk in food and nutritional security to the increasing population.

Burgeoning growth of populations in the last several centuries has resulted in demands for sustainable food production, processing technologies and its industrialization, which is the main source of food and the backbone of economies. Therefore, bridging the balance is required between the demand of sustainable food production, processing, supply, and maintaining socio-economically healthy environment for long-term survival of human life, which will be the most important challenge for humankind in the coming decades. To surmount this problem and to meet the growing demands for food supplies, new tools of science, novel processes and emerging technologies are being developed and create tremendous opportunities for agricultural applications. And the new technologies, mechanization, use of chemicals, specialization and various governmental policies have increased food and crop plants productivity and have resulted in favoring enhancement of storage and economy.

The book, *Engineering Interventions in Foods and Plants,* focuses on these issues, it evaluates the food supply chain, provides detailed descriptions of food production and processing using the tools of life cycle assessment, presents improved technological approaches for agricultural practices, explores sustainable processing techniques, discusses emerging analytical techniques for research and development, looks at the challenges associated with the use of agricultural resources to grow biofuels and bio-based products, presents technologies for the reduction of process-induced toxins generated, considers social factors influencing consumer perceptions about current and emerging technologies, and explains the importance of maintaining biodiversity in sustainable human diets. Therefore, this book volume will be an asset to improve and to provide knowledge and feedback in the progress and future opportunities for all those who are associated with agricultural and biological engineering.

Part One of this book, *Emerging Entrepreneurship for Rural Areas,* consists of three chapters ("Beekeeping Technology and Honey Processing: Emerging Entrepreneurship for Rural Areas"; "Herbal Formulations for Treatment of Dental Diseases: Perspectives, Potential, and Applications"; and "Engineering Interventions for Extraction of Essential Oils from Plants") devoted to entrepreneur opportunities for rural peoples with employment of food and agricultural processing.

Part Two, *Engineering Interventions in Health Benefits,* contains three chapters ("Processing Technology and Potential Health Benefits of Coffee"; "Biochemical Composition, Processing Technology and Health Benefits of Green Tea: A Review"; and "Effects of Thermal Processing on Nutritional Composition of Green Leafy Vegetables: A Review") dealing with research opportunities and novel practices in human health. These three chapters are devoted on biochemical composition, processing technology, and human health benefits.

Part Three, *Management Strategies in Agricultural Engineering,* consists of two chapters. Chapter 7 ("Ethanol Production from Different Substrates: Effects on Environmental Factors and Potential Applications") describes production technology of ethanol from sources and its potential application in various industries like chemical, food, pharmaceutical as well as biofuel. And the last chapter ("Food Grain Storage Structures: Introduction and Applications") focuses on minimizing the losses from microbial issues as well as from insect-pest problems during grain storage by different storage structures to fulfill an increasing demand of rapidly growing populations of the world.

With contributions from a broad range of leading professors and scientists, this book focuses on areas of processing technologies in foods and plants. It will provide a guide to students, instructors and researchers of foods and plants. In addition, food and plant science professionals who are seeking recent advanced and innovative knowledge in processing will find this book helpful. It is envisaged that this book will also serve as a reference source for individuals engaged in research, processing and product development in foods and plant science areas.

With great pleasure, I would like to extend my sincere thanks to all the learned contributors for excellent contributions and attention details and accuracy of information presented as well as their consistent support and cooperation. They have made our task as editors a pleasure. I also thank Megh R. Goyal, Senior Editor-in-Chief, for his guidance, mission, and vision.

It is hoped that this edition will stimulate discussion and generate helpful comments to improve upon future editions. Efforts are made to cross-reference the chapters as such.

Finally, we acknowledge Almighty God, who provided all the inspirations, insights, positive thoughts, and channels to complete this book project.

—Deepak Kumar Verma
Editor

PREFACE 2 BY MEGH R. GOYAL

This book, *Engineering Interventions in Foods and Plants,* is unique because it encompasses new technologies of beekeeping, extraction of essential oils from plants, potential of herbal formulations to cure dental conditions and diseases, potential of coffee/tea, and poiential of leafy vegetables to keep us healthy.

The weblink "https://en.wikipedia.org/wiki/Beekeeping" indicates that *"Apiculture (from Latin: apis, bee) is the maintenance of honey bee colonies, commonly in hives, by humans. A beekeeper (or apiarist) keeps bees in order to collect their honey and other products that the hive produces (including beeswax, propolis, pollen, and royal jelly), to pollinate crops, or to produce bees for sale to other beekeepers. A location where bees are kept is called an apiary or bee yard."* The Orange Blossom Beekeepers Association (OBBA) is a collective of commercial and vocational beekeepers from across Central Florida. The club strives to mentor novices and to serve as a resource for experienced beekeepers. In addition, OBBA (http://www.orangeblossombeekeepers.org/) seeks to educate the public about the importance of bees and beekeeping. Throughout the year, OBBA conducts interactive workshops in its apiary and brings in expert speakers for monthly club meetings. Also, OBBA attends and hosts educational events throughout Central Florida. I invite readers to have a taste of beekeeping in this book volume.

Apple Academic Press, Inc., published my first book on *Management of Drip/Trickle or Micro Irrigation*, a 10-volume set under the book series *Research Advances in Sustainable Micro Irrigation*, in addition to other books in the focus areas of agricultural and biological engineering. The mission of this book series is to introduce the profession of agricultural and biological engineering. I cannot guarantee the information in this book series will be enough for all situations.

At 49th Annual Meeting of the Indian Society of Agricultural Engineers at Punjab Agricultural University during February 22–25 of 2015, a group of ABEs convinced me that there is a dire need to publish book volumes on the focus areas of agricultural and biological engineering (ABE). This is how the idea was born to create this new book series titled, *Innovations in Agricultural and Biological Engineering.*

The contributions by all cooperating authors to this book volume have been most valuable in the compilation. Their names are mentioned in each chapter and in the list of contributors. This book would not have been written without the valuable cooperation of these investigators, many of whom are renowned scientists who have worked in the field of ABE throughout their professional careers. Dr. Deepak Kumar Verma joins me as a Editor of this book volume. Dr. Verma is a frequent contributor to my book series and a staunch supporter of my profession. His contribution to the content and quality of this book has been invaluable.

I want to thank editorial staff, Sandy Jones Sickels, Vice President, and Ashish Kumar, Publisher and President at Apple Academic Press, Inc., for making every effort to publish the book when diminishing water resources are a major issue worldwide. Special thanks are due to the AAP Production Staff for typesetting for the quality production of this book.

I request readers to offer constructive suggestions that may help to improve the next edition.

I express my deep admiration to my family for their understanding and collaboration during the preparation of this book.

As an educator, there is a piece of advice to one and all in the world: *"Permit that our Almighty God, our Creator, allow us to inherit new technologies for a better life at our planet. I invite my community in agricultural engineering to contribute book chapters to the book series by getting married to my profession".* I am in total love with our profession by length, width, height and depth. Do you?

—*Megh R. Goyal, PhD, PE*
Senior Editor-in-Chief

WARNING/DISCLAIMER

PLEASE READ CAREFULLY

The goal of this book volume, *Engineering Interventions in Foods and Plants,* is to guide the world community on how make efficient use of technology available for different processes in food science and technology. The reader must be aware that the dedication, commitment, honesty, and sincerity are important factors for success.

The editors, the contributing authors, the publisher, and the printer have made every effort to make this book as complete and as accurate as possible. However, there still may be grammatical errors or mistakes in the content or typography. Therefore, the content in this book should be considered as a general guide and not a complete solution to address any specific situation in food engineering. For example, one type of food process technology does not fit all case studies in dairy engineering/science/technology.

The editors, the contributing authors, the publisher, and the printer shall have neither liability nor responsibility to any person, any organization or entity with respect to any loss or damage caused, or alleged to have caused, directly or indirectly, by information or advice contained in this book. Therefore, the purchaser/reader must assume full responsibility for the use of the book or the information therein.

The mention of commercial brands and trade names are only for technical purposes. A particular product is not endorsed over another product or equipment not mentioned. The editors, cooperating authors, educational institutions, and the publisher Apple Academic Press Inc. do not have any preference for a particular product.

ABOUT THE EDITOR

Deepak Kumar Verma, PhD
Research Scholar, Department of Agricultural and Food Engineering, Indian Institute of Technology, West Bengal, India

Deepak Kumar Verma is an agricultural science professional and is currently a PhD Research Scholar in the specialization of food processing engineering in the Agricultural and Food Engineering Department, Indian Institute of Technology, Kharagpur (WB), India. In 2012, he received a DST-INSPIRE Fellowship for PhD study by the Department of Science & Technology (DST), Ministry of Science and Technology, Government of India. Mr. Verma is currently working on the research project "Isolation and Characterization of Aroma Volatile and Flavoring Compounds from Aromatic and Non-Aromatic Rice Cultivars of India." His previous research work included "Physico-Chemical and Cooking Characteristics of Azad Basmati (CSAR 839-3): A Newly Evolved Variety of Basmati Rice (*Oryza sativa* L.)". Apart from his area of specialization in plant biochemistry, he has also built a sound background in plant physiology, microbiology, plant pathology, genetics and plant breeding, plant biotechnology and genetic engineering, seed science and technology, food science and technology, etc. In addition, he is a member of different professional bodies, and his activities and accomplishments include conferences, seminars, workshop, training, and also the publication of research articles, books, and book chapters. He earned his BSc degree in agricultural science from the Faculty of Agriculture at Gorakhpur University, Gorakhpur, and his MSc (Agriculture) in Agricultural Biochemistry in 2011. He also received an award from the Department of Agricultural Biochemistry, Chandra Shekhar Azad University of Agricultural and Technology, Kanpur, India. Readers may contact him at: rajadkv@rediffmail.com or deepak.verma@agfe.iitkgp.ernet.in.

ABOUT THE SENIOR EDITOR-IN-CHIEF

 **Megh R. Goyal, PhD**
*Retired Professor in Agricultural and Biomedical
Engineering, University of Puerto Rico,
Mayaguez Campus Senior Acquisitions Editor,
Biomedical Engineering and Agricultural Science,
Apple Academic Press, Inc.
E-mail: goyalmegh@gmail.com*

Megh R. Goyal, PhD, PE, is a Retired Professor in Agricultural and Biomedical Engineering from the General Engineering Department in the
College of Engineering at University of Puerto Rico–Mayaguez Campus; and Senior Acquisitions Editor and Senior Technical Editor-in-Chief
in Agriculture and Biomedical Engineering for Apple Academic Press
Inc. He has worked as a Soil Conservation Inspector and as a Research
Assistant at Haryana Agricultural University and Ohio State University.
He was the first agricultural engineer to receive the professional license
in Agricultural Engineering in 1986 from the College of Engineers and
Surveyors of Puerto Rico. On September 16, 2005, he was proclaimed
as "Father of Irrigation Engineering in Puerto Rico for the twentieth century" by the ASABE, Puerto Rico Section, for his pioneering work on
micro irrigation, evapotranspiration, agroclimatology, and soil and water
engineering. During his professional career of 45 years, he has received
many prestigious awards and honors, including being recognized as one
of the experts "who rendered meritorious service for the development of
[the] irrigation sector in India" by the Water Technology Centre of Tamil
Nadu Agricultural University in Coimbatore, India. A prolific author and
editor, he has written more than 200 journal articles and textbooks and
has edited over 50 books. He received his BSc degree in engineering
from Punjab Agricultural University, Ludhiana, India; his MSc and PhD
degrees from Ohio State University, Columbus; and his Master of Divin-
ity degree from Puerto Rico Evangelical Seminary, Hato Rey, Puerto
Rico, USA. Readers may contact him at: goyalmegh@gmail.com.

BOOKS ON AGRICULTURAL AND BIOLOGICAL ENGINEERING FROM APPLE ACADEMIC PRESS, INC.

Management of Drip/Trickle or Micro Irrigation
Megh R. Goyal, PhD, PE, Senior Editor-in-Chief

Evapotranspiration: Principles and Applications for Water Management
Megh R. Goyal, PhD, PE, and Eric W. Harmsen, Editors

Book Series: Research Advances in Sustainable Micro Irrigation
Senior Editor-in-Chief: Megh R. Goyal, PhD, PE
  Volume 1: Sustainable Micro Irrigation: Principles and Practices
  Volume 2: Sustainable Practices in Surface and Subsurface Micro Irrigation
  Volume 3: Sustainable Micro Irrigation Management for Trees and Vines
  Volume 4: Management, Performance, and Applications of Micro Irrigation Systems
  Volume 5: Applications of Furrow and Micro Irrigation in Arid and Semi-Arid Regions
  Volume 6: Best Management Practices for Drip Irrigated Crops
  Volume 7: Closed Circuit Micro Irrigation Design: Theory and Applications
  Volume 8: Wastewater Management for Irrigation: Principles and Practices
  Volume 9: Water and Fertigation Management in Micro Irrigation
Volume 10: Innovation in Micro Irrigation Technology

Book Series: Innovations and Challenges in Micro Irrigation
Senior Editor-in-Chief: Megh R. Goyal, PhD, PE
Volume 1: Principles and Management of Clogging in Micro Irrigation
Volume 2: Sustainable Micro Irrigation Design Systems for Agricultural Crops: Methods and Practices
Volume 3: Performance Evaluation of Micro Irrigation Management: Principles and Practices
Volume 4: Potential Use of Solar Energy and Emerging Technologies in Micro Irrigation
Volume 5: Micro Irrigation Management: Technological Advances and Their Applications

Volume 6: Micro Irrigation Engineering for Horticultural Crops: Policy Options, Scheduling, and Design
Volume 7: Micro Irrigation Scheduling and Practices
Volume 8: Engineering Interventions in Sustainable Trickle Irrigation Water Requirements, Uniformity, Fertigation, and Crop Performance

Book Series: Innovations in Agricultural and Biological Engineering
Senior Editor-in-Chief: Megh R. Goyal, PhD, PE
- Dairy Engineering: Advanced Technologies and their Applications
- Developing Technologies in Food Science: Status, Applications, and Challenges
- Emerging Technologies in Agricultural Engineering
- Engineering Interventions in Agricultural Processing
- Engineering Interventions in Foods and Plants
- Engineering Practices for Agricultural Production and Water Conservation: An Interdisciplinary Approach
- Flood Assessment: Modeling and Parameterization
- Food Engineering: Modeling, Emerging Issues, and Applications.
- Food Process Engineering: Emerging Trends in Research and Their Applications
- Food Technology: Applied Research and Production Techniques
- Modeling Methods and Practices in Soil and Water Engineering
- Novel Dairy Processing Technologies: Techniques, Management, and Energy Conservation
- Processing Technologies for Milk and Milk Products: Methods, Applications, and Energy Usage
- Soil and Water Engineering: Principles and Applications of Modeling
- Soil Salinity Management in Agriculture: Technological Advances and Applications
- Technological Interventions in Dairy Science: Innovative Approaches in Processing, Preservation, and Analysis of Milk Products
- Technological Interventions in Management of Irrigated Agriculture
- Technological Interventions in the Processing of Fruits and Vegetables
- Scientific and Technical Terms in Bioengineering and Biological Engineering
- State-of-the-Art Technologies in Food Science: Human Health, Emerging Issues and Specialty Topics
- Sustainable Biological Systems for Agriculture: Emerging Issues in Nanotechnology, Biofertilizers, Wastewater, and Farm Machines

EDITORIAL

Under the book series titled *Innovations in Agricultural and Biological Engineering*, Apple Academic Press, Inc., (AAP) is publishing book volumes in specialty areas over a span of 8 to 10 years. These specialty areas have been defined by the *American Society of Agricultural and Biological Engineers* (http://asabe.org). AAP wants to be the principal source of books in the field of agricultural and biological engineering. We seek book proposals from the readers in area of their expertise.

The mission of this series is to provide knowledge and techniques for agricultural and biological engineers (ABEs). The series aims to offer high-quality reference and academic content in agricultural and biological engineering (ABE) that is accessible to academicians, researchers, scientists, university faculty, and university-level students and professionals around the world. The following material has been edited/modified and reproduced below *"Goyal, Megh R., 2006. Agricultural and biomedical engineering: Scope and opportunities. Paper Edu_47 at the Fourth LACCEI International Latin American and Caribbean Conference for Engineering and Technology (LACCEI' 2006): Breaking Frontiers and Barriers in Engineering: Education and Research by LACCEI University of Puerto Rico – Mayaguez Campus, Mayaguez, Puerto Rico, June 21–23."*

WHAT IS AGRICULTURAL AND BIOLOGICAL ENGINEERING (ABE)?

"Agricultural Engineering (AE) involves application of engineering to production, processing, preservation and handling of food, fiber, and shelter. It also includes transfer of technology for the development and welfare of rural communities," according to http://isae.in." *ABE is the discipline of engineering that applies engineering principles and the fundamental concepts of biology to agricultural and biological systems and tools, for the safe, efficient and environmentally sensitive production, processing,*

and management of agricultural, biological, food, and natural resources systems," according to http://asabe.org.

"AE is the branch of engineering involved with the design of farm machinery, with soil management, land development, and mechanization and automation of livestock farming, and with the efficient planting, harvesting, storage, and processing of farm commodities," definition by: http://dictionary.reference.com/browse/agricultural+engineering.

"AE incorporates many science disciplines and technology practices to the efficient production and processing of food, feed, fiber and fuels. It involves disciplines like mechanical engineering (agricultural machinery and automated machine systems), soil science (crop nutrient and fertilization, etc.), environmental sciences (drainage and irrigation), plant biology (seeding and plant growth management), animal science (farm animals and housing), etc.," by: http://www.ABE.ncsu.edu/academic/agricultural-engineering.php.

"According to https://en.wikipedia.org/wiki/Biological_engineering: *"BE (Biological engineering) is a science-based discipline that applies concepts and methods of biology to solve real-world problems related to the life sciences or the application thereof. In this context, while traditional engineering applies physical and mathematical sciences to analyze, design and manufacture inanimate tools, structures and processes, biological engineering uses biology to study and advance applications of living systems."*

SPECIALTY AREAS OF ABE

Agricultural and Biological Engineers (ABEs) ensure that the world has the necessities of life including safe and plentiful food, clean air and water, renewable fuel and energy, safe working conditions, and a healthy environment by employing knowledge and expertise of sciences, both pure and applied, and engineering principles. Biological engineering applies engineering practices to problems and opportunities presented by living things and the natural environment in agriculture. BA engineers understand the interrelationships between technology and living systems, have available a wide variety of employment options. *"ABE embraces a variety of following specialty areas,"* http://asabe.org. As new technology and information emerge, specialty areas are created, and many overlap with one or more other areas.

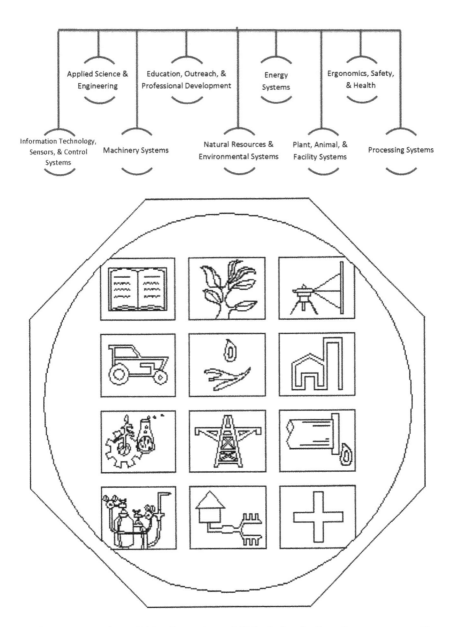

1. **Aquacultural Engineering**: ABEs help design farm systems for raising fish and shellfish, as well as ornamental and bait fish. They specialize in water quality, biotechnology, machinery, natural resources, feeding and ventilation systems, and sanitation. They

seek ways to reduce pollution from aquacultural discharges, to reduce excess water use, and to improve farm systems. They also work with aquatic animal harvesting, sorting, and processing.

2. **Biological Engineering** applies engineering practices to problems and opportunities presented by living things and the natural environment.

3. **Energy:** ABEs identify and develop viable energy sources – biomass, methane, and vegetable oil, to name a few – and to make these and other systems cleaner and more efficient. These specialists also develop energy conservation strategies to reduce costs and protect the environment, and they design traditional and alternative energy systems to meet the needs of agricultural operations.

4. **Farm Machinery and Power Engineering**: ABEs in this specialty focus on designing advanced equipment, making it more efficient and less demanding of our natural resources. They develop equipment for food processing, highly precise crop spraying, agricultural commodity and waste transport, and turf and landscape maintenance, as well as equipment for such specialized tasks as removing seaweed from beaches. This is in addition to the tractors, tillage equipment, irrigation equipment, and harvest equipment that have done so much to reduce the drudgery of farming.

5. **Food and Process Engineering:** Food and process engineers combine design expertise with manufacturing methods to develop economical and responsible processing solutions for industry. Also food and process engineers look for ways to reduce waste by devising alternatives for treatment, disposal and utilization.

6. **Forest Engineering**: ABEs apply engineering to solve natural resource and environment problems in forest production systems and related manufacturing industries. Engineering skills and expertise are needed to address problems related to equipment design and manufacturing, forest access systems design and construction; machine-soil interaction and erosion control; forest operations analysis and improvement; decision modeling; and wood product design and manufacturing.

7. **Information and Electrical Technologies Engineering** is one of the most versatile areas of the ABE specialty areas, because it is

applied to virtually all the others, from machinery design to soil testing to food quality and safety control. Geographic information systems, global positioning systems, machine instrumentation and controls, electromagnetics, bioinformatics, biorobotics, machine vision, sensors, spectroscopy: These are some of the exciting information and electrical technologies being used today and being developed for the future.

8. **Natural Resources:** ABEs with environmental expertise work to better understand the complex mechanics of these resources, so that they can be used efficiently and without degradation. ABEs determine crop water requirements and design irrigation systems. They are experts in agricultural hydrology principles, such as controlling drainage, and they implement ways to control soil erosion and study the environmental effects of sediment on stream quality. Natural resources engineers design, build, operate and maintain water control structures for reservoirs, floodways and channels. They also work on water treatment systems, wetlands protection, and other water issues.

9. **Nursery and Greenhouse Engineering**: In many ways, nursery and greenhouse operations are microcosms of large-scale production agriculture, with many similar needs – irrigation, mechanization, disease and pest control, and nutrient application. However, other engineering needs also present themselves in nursery and greenhouse operations: equipment for transplantation; control systems for temperature, humidity, and ventilation; and plant biology issues, such as hydroponics, tissue culture, and seedling propagation methods. And sometimes the challenges are extraterrestrial: ABEs at NASA are designing greenhouse systems to support a manned expedition to Mars!

10. **Safety and Health:** ABEs analyze health and injury data, the use and possible misuse of machines, and equipment compliance with standards and regulation. They constantly look for ways in which the safety of equipment, materials and agricultural practices can be improved and for ways in which safety and health issues can be communicated to the public.

11. **Structures and Environment:** ABEs with expertise in structures and environment design animal housing, storage structures, and

greenhouses, with ventilation systems, temperature and humidity controls, and structural strength appropriate for their climate and purpose. They also devise better practices and systems for storing, recovering, reusing, and transporting waste products.

CAREER IN AGRICULTURAL AND BIOLOGICAL ENGINEERING

One will find that university ABE programs have many names, such as biological systems engineering, bioresource engineering, environmental engineering, forest engineering, or food and process engineering. Whatever the title, the typical curriculum begins with courses in writing, social sciences, and economics, along with mathematics (calculus and statistics), chemistry, physics, and biology. Student gains a fundamental knowledge of the life sciences and how biological systems interact with their environment. One also takes engineering courses, such as thermodynamics, mechanics, instrumentation and controls, electronics and electrical circuits, and engineering design. Then student adds courses related to particular interests, perhaps including mechanization, soil and water resource management, food and process engineering, industrial microbiology, biological engineering or pest management. As seniors, engineering students team up to design, build, and test new processes or products.

For more information on this series, readers may contact:

Ashish Kumar, Publisher and President	Megh R. Goyal, PhD, PE
Sandy Sickels, Vice President	Book Series Senior
Apple Academic Press, Inc.	Editor-in-Chief
Fax: 866-222-9549	*Innovations in Agricultural*
E-mail: ashish@appleacademicpress.com	*and Biological Engineering*
http://www.appleacademicpress.com/	E-mail: goyalmegh@gmail.
publishwithus.php	com

PART I

EMERGING ENTREPRENEURSHIP
FOR RURAL AREAS

CHAPTER 1

BEEKEEPING TECHNOLOGY AND HONEY PROCESSING: EMERGING ENTREPRENEURSHIP FOR RURAL AREAS

VISHAL SINGH, DEEPAK KUMAR VERMA, and DEEPALI CHAUHAN

CONTENTS

1.1 INTRODUCTION

Beekeeping emerged as an economical and profitable culture in 19th century due to implementation of advance and scientific approaches and generally practiced in places like forest areas, hills and agricultural lands, etc.

Movable frame hives were constructed initially in 1880 A.D. and practiced for keeping Bees (*Apis cerana* F.) in Bengal [49]. Within last decades of 21st century, it has been observed that honeybee colonies are decreased continuously due to reduction in apicultural activities which draw the more intension of scientific research to enhance the beekeeping practices [14]. Aizen et al. [4] have observed that in last five decades crop production was not affected due to decreasing of plants associated with honeybee colony as well as studied that grown of pollinator associated crops also increased for catalyzed the beekeeping [3, 13].

Collected nectar by bees mainly contains: glucose (24–40%), fructose (30–45%) and sucrose (0.1–4.8%) like carbohydrates [53]. Apiculture is bringing attention as important agricultural occupation due to the importance of its products like honey, beeswaxes, pollen, etc. Enhancement in production of apiculture has been focused on production and quality maintenance throughout the process of beehives design to honey collection and further processing requires intensive scientific techniques. Collected nectar changed in honey with the help of honeybees, which carried different constituents like carbohydrate, water, minerals and enzymes, etc. in which some essential nutrients contributed and some transported from the plants by honeybees [5].

Adequate management practices, optimum knowledge of beekeeping like nature of honeybees, optimum climatic condition and suitable places for beekeeping have increased the quality as well as quantity of apicultural produces and also have minimized the unwanted risk. Beekeeping is growing as beneficial, eco-friendly and economically agricultural practice but still beekeeping is more practiced in rural or interior part of our country and requires more awareness and support from the government agencies.

India have different regions, which are known for apicultural potential like Himalayan forest area, Jammu and Kashmir hilly area, north-eastern and west region, forest of western and eastern Ghats, crop land of pulses, legumes and oil seed, etc. [52]. In winter conditions, low temperature is not suitable for honeybee colonies growth reported in several countries of the world like Middle East and Europe, etc. [36, 43]. Beekeeping has been performed since several hundred years for getting nutritious product like honey and commercial produces like beeswax. Honeybees gather nectar from the flowers and other plants and the nectar is stored in its house called beehives and finally a honey is produced with sweet flavored as well as attractive aroma, which should be preserved during extraction and further processing steps [51].

This chapter discusses the potential of beekeeping and honey technology in India. Principles and applications are presented.

1.2 BEEKEEPING AND HONEY: BRIEF OUTLOOK

1.2.1 BEEKEEPING: AT A GLANCE

Maintenance of honeybee colonies and honey production is known as beekeeping or Apiculture. In Apiculture, honeybees stay in hives (it may be either artificial designed by bee-keeper or natural made by bees in trees and buildings) but commonly farmers are taking help of artificial beehives to keep the honeybee for production of honey. Bees are allowed to collect the sugar containing nectars known as honey, from nearby flowers, trees and crops through pollination and stored it in their hives. Beekeepers have to collect the major amount honey (some amount of honey should be left in hives for the bees consumption), at regular intervals. After extraction of honey from hives, it requires further filtration and processing for making is consumable for avoiding the different kinds of impurities and contamination due to insects and ants, etc.

1.2.2 HONEY: AT A GLANCE

Honey extraction is performed with the help of local available and cheap methods like bamboo strips of length between 2–2.5 meters depending upon the beehives position tied just beneath the bee-comb, which provides support to the rest part of bee-comb after separation of honey and also helps to rebuilt the comb for repeated collection of honey performed from single colony. Honey collected from the combs with the help of container is attached with screen without any impurities and other produces like wax, pollen, etc. also collected without any damage. Today, farmers and entrepreneurs are interested in apiculture due to its demand and want to earn higher profit by getting large production and collection of honey with the help of advance, scientific and technical methods of honey collection. Honey contains carbohydrate (82%), moisture (17%), vitamins (like Vitamins B and C) and antioxidants (like flavonoids) [29]. For proper health benefits, adequate amount of honey (2–4 table spoons) should be consumed because excess doses are not beneficial for the human body [10]. Beekeeping is performed in different regions of India on plants [52], which are frequently available according to the climatic suitability of that region (Table 1.1). Burlando et al. [12] stated that honey is used as food commodities simultaneously and is also used for several pharmaceutical purposes (like for preparation of medicines and syrup, etc.); and for cosmetic manufacturing (like lotions, soaps and cream, etc.). Honey is utilized as a rich source of carbohydrate in food, which is beneficial for human health especially for brain development [15].

TABLE 1.1   Different Types of Plants for Apiculture Farming

Types of plants used for beekeeping	Forest/Region
Citrus	Himachal Pradesh
Coffee	Karnataka
Coriander	West Bengal
Jamun	Mahabaleswar
Litchi	Bihar
Mohul	Madhya Pradesh
Prosopis	Kutch
Rubber	Kerala
Sandal wood	Kadappa forest
Soapnut	Coastal Karnataka and Andhra Pradesh
Sula	Himalayan region

1.3   EQUIPMENT AND HONEYBEE SPECIES USED FOR BEEKEEPING

1.3.1   EQUIPMENT USED FOR BEEKEEPING

Generally, traditional apiculture performs on a small-scale beekeeping for earning livelihood and for producing the honey at small-scale. Simple and local constructed equipments are being used but scientific equipments are based on advance technology like honey quality testing gauze and filtering gauze, etc. that are required for apiculture at large scale [6]. With the intensive emphases, apiculture opens wide scope for beekeeping as profitable culture at small as well as large scale and it needs different equipments for easy, safe, least contaminated and heavy production in less time. For constructing adequate hives, one needs hive stand, hive bodies, honey supers, top cover, inner cover and bottom board.

1.3.1.1   Hive Tools

Hive tools are commonly constructed with steel and sharpened at one end (appears same as screw driver) or one side (appears same as knife) and opposite side is used for hand holding. Hive tool helps to separate the beeswaxes by removing the unwanted part of it, separating from frame and supports to each-other and separate or removing of other portions of beeswax [6].

1.3.1.2   Bees Hives

Beehives is defined as a natural or artificial structure in which bees are staying, storing honey and multiplying (it is a house for the honeybees). Generally, honeybee forms their hives or colonies by themselves in a trees (trees which bear flowers and fruits) and buildings nearer to flowers and fruits bearing crops for easy pollination. Natural beehives looks like a hanging bird nest. Artificial hives are made with wood in different rectangular segments, which are arranged in a sequence within a wooden box for keeping the domestic bees with the purpose of honey production. Only the western honeybee and the eastern honeybee are *A. mellifera*) and *A. cerana*, respectively are domesticated by humans.

1.3.1.2.1   Types of Bees Hives

Traditional Hives: Traditional hives are constructed in local area where beekeeping takes place at small level and material for making hives should be available in rural areas like clay containers, grasses and hollow logs, etc. [1] and constructed traditional hives facilitate the bees to stay within hive in such a way that beekeeper can easily extract the bees produce but in traditional method possibilities of death of many bees during collection of honey and beeswaxes.

Intermediate Technology Hives: Intermediate technology hive is the result of combination of both traditional and movable frame hives because generally available materials in rural places are used for constructing this hives [1]. For proper building the bee combs, container is attached with number of strips, which facilitate the beekeeper to easily remove and separate bee combs [6].

Honeycombs from Traditional Hives: Combs contain pure and natural honey because it does not come in the contact with air or humans that is why honey in these combs is more qualitative as well as more flavored than extracted and processed honey. For natural appearance, honeycombs in the form of separated pieces of combs full with honey can be kept carefully for marketable value [6].

Movable Frame Hives: Movable frame hives are manufactured with wood or plastic frame, which is inserted within a container or box and number of boxes are inserted into single hive one above the other for creating more spaces for honey storage. It is very popular and frequently used in regions, where apiculture activities are considered as major and important agricultural business because movable frame hive facilitates the more extraction of apicultural produce like honey and beeswaxes, etc. Up to long duration of seasons without any destruction in the natural activities also facilitate the beekeeper to proper inspection and growth of the bee comb [6].

Honeycombs from Frame Hives: for obtaining honey from frame hives, thin and less strong wires are considered than the wired frames and it is easy to separate honeycombs, which contain fresh and pure honey [2].

1.3.1.2.2 *Parts of Bees Hives*

- *The Bottom Board* is used to provide safety from bottom side and stability to the hive. Design of bottom board depends on the place of hives set up and desired height from ground surface. Bottom board is used to prevent the hive to avoid damage of bottom parts of hive due to moisture, which primarily degrades the bottom board. For minimizing the any deterioration from moisture, 5–6 inches elevation is required [2].

- **Inner and Top Cover** is generally arranged in systematically for capping the hive to avoid the destruction of hives due to high speed wind and water drops. A small opening in the top cover is provided for easy ventilation, which is very essential in winter season for avoiding the chance of condensation within the top cover and hives. Inner cover is arranged in such a way that nearly 0.25 inches space should be available for free movement of bees and dry sugar is also supplied to the bees during winter in the case of less food were stored. Inner cover facilitates sufficient ventilation during summer.

- **Honey Super:** Honey supers have different sizes depending on the size of hives. Larger honey supers are used in hives, which can have more amount of honey and generally are placed on the top of the hives to facilitate the proper honey collection. Different sizes of honey super can be made such as: small, medium and larger size super.

1.3.2 HONEYBEE SPECIES FOR BEEKEEPING IN INDIA

There are five different types of honeybee found in India: *Apis cerena; Apis dorsata; Apis florea; Apis mellifera; and Trigona* spp. **Scientific classifications of** different species of honeybee have been indicated by many investigators [18, 21, 31, 38].

1.3.2.1 *Apis dorsata* (Rock bee or giant bee)

Apis dorsata is widely present in the forest and hilly areas of Northern region (basically nearby Himalayan range) and Central regions (Madhya Pradesh and Chhattisgarh), Sundarban region of West-Bengal and southern regions (Nilgiri hills and Western Ghats, etc.). The local people of these regions collect the main products (honey) and other side products (bee wax, milk, etc.) in large quantities for earning their livelihood by selling it in the local market. *A. dorsata*, the organic honeybees from the Sunderban forests in West Bengal, are of great demand today. The southern part of India is also having large number of *A. dorsata* colonies and contribute large share of total honey production in India.

1.3.2.2   *Apis florea* (Little bee)

Apis florea commonly exists in central, desert (e.g., some parts of Rajasthan and Gujarat), plane (e.g., Bihar, Uttar Pradesh and Madhya Pradesh, etc.) and Forest (e.g., Odisha, Uttarakhand, Malabar coast, etc.) regions of India. Soman and Chawda [50] observed that Kutch region of Gujarat is a place, where plenty of *A. florae* colonies are present, which contribute a major portion of honey production.

1.3.2.3   *Trigona* spp. (Dammar bee)

Trigona spp. is found commonly in all parts of the country and remains long periods in the same abode. Dammer bee is a small size species usually collecting nectars from flowers and trees. Honey produced by Dammer bees are produced in small quantity that is why it consumed by local people and available in local market only as well as is used in preparation of local and traditional medicine.

1.3.2.4   *Apis mellifera*

Apis mellifera is a popular species of bees and is widely used for honey production throughout the world. In India, initially honey production by *A. mellifera* was started with help of human made beehives in Punjab. Due to good honey yield and domestic nature, it has become popular in entire India (e.g., Himachal Pradesh, Bihar, Uttar-Pradesh, West-Bengal, Maharashtra, Karnataka, Kerala, and Jharkhand). Some important characteristics of *A. mellifera* are shown in Table 1.2.

TABLE 1.2 Important Characteristics of *Apis* spp. viz. *A. cerana* **vs** *A. mellifera*

Salient characteristic	A. cerana	A. mellifera	Reference
Active foraging time and temperature range	9 am and 11.30 am/15.5–21°C	11 am and 1 pm/21–25°C	[33]
Body mass	43.8 mm	77.2 mm	[16]
Collecting propolis	No	Yes	[19]
Comb structure	Uneven round	Square	[40]
Defense to wasps	Forming a ball with worker	Individual	[41, 55]
Dismantling of old comb	Yes	No	[19]
Flight pattern	Rapid, hasty, and unpredictably zig-zag	Steady and clumsy	[19]
Flower preference	Wide range of flowers, including wild plants	Mainly on Trifolium and Brasscia	[35]
Foraging range	Maximum 1500 m to 2500 m	Average 1650 m, maximum 6 km	[32]
Frequencies of absconding	Often (heavily depends on condition)	Rarely (heavily depends on condition)	[17]
Homing speed (~50 m)	192 sec	295 sec	[8]
Nest cavity volume	10–15 liters in general	35 liters	[47]
Origin	Southeast Asia	Africa or Western Asia	[40]
Pollination	Larger pollen collector	Smaller pollen collector (compared to *A. cerana*)	[7]
Pore in the drone cell	Yes	No	[23]
Rate of stinging	Low	High	[19]
Robbing defense	Weak	Strong	[7]
Royal jelly production rate	3.21 ± 0.43 g	80.5 ± 7.83 g	[33]
Varroa mite resistance	Yes	No	[39]
Ventilation direction	Head toward outside	Head toward entrance	[38, 55]
Wing beat frequencies (worker)	306 beats/s	235 beats/s	[40]
Wing length (worker)	7.54 ± 1.14 mm	9.32 ± 0.16 mm	[48]

1.3.2.5 *Apis cerena* indica

Apis cerena indica can be found in all parts of India. They make their honeycomb in closed places, which are away from light. These are not suited for artificial beekeeping. Some important characteristics of *A. cerana* are depicted in Table 3.

1.4 OTHER IMPORTANT TOOLS USED IN BEEKEEPING

- *Smoker* is used to supply smokes near the mouth of hive (the place from where hive opens first) very carefully and in adequate quantity. Due to the smoke, the honeybees are gathered and tried to stick with honey. It calms the bees effectively and facilitates the honey extraction. Generally smoker assembled with different parts (like fuel chamber, fuel combustion chamber, operating handle and smoke driven nozzle, etc., cotton waste, burlap, wood, animal like cat hair, etc.) is used as burning material after supplying the fire which responsible for producing the smoke (Figure 1.1A).
- *Mesh Helmet* is used for covering the human head and neck to protect the entire face before performing any activity with beehive (Figure 1.1B). Veils made with wire can protect the face and other portion of body that has been covered with clothes. Light color should be used because it is observed that bees attack on dark colored clothes.
- *Electric Uncapping Knife* is used for removing the wax cap from the beehives before extraction of honey and is also used for proper cleaning of hives and its parts. For performing above mentioned operations efficiently and quickly, the electrical operated knives are being used rather than the manual knife (Figure 1.1C).
- *Hive Stand* provides proper stability and platform height to the hive but generally hive constructed in rural areas has no stand for reducing the construction cost as well as easy hive portability (Figure 1.1D).
- *Bee brush* is equipment used for cleaning the hive frames and also to separate honeybees from the frame gently. Generally bee brush is operated manually (Figure 1.1E).
- *Sting Resistant Gloves* protect the hands from honeybee sting during handling of beehives (Figure 1.1F). Gloves should be made with flexible clothes or leather, which facilitates the easy movement to the hand and fingers [30].

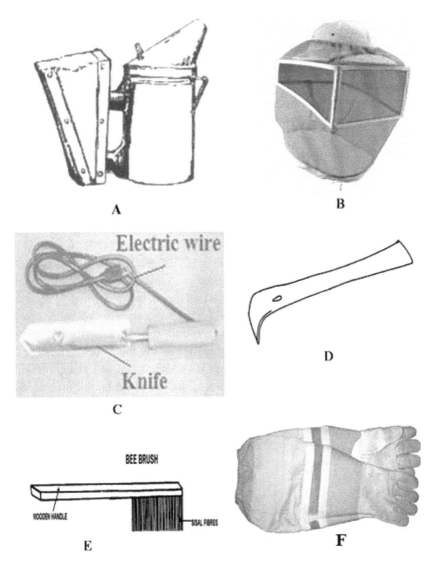

FIGURE 1.1 Important tools of beekeeping. (A) Bees smoker, (B) Mesh helmet, (C) Electric uncapping knife, (D) Hive tool, (E) Bee brush, and (F) Sting resistant gloves.

1.5 EQUIPMENTS FOR COLLECTING HONEYBEE'S PRODUCTS

- ***Honey Extractor*** is a rotating device of cylindrical shape structure commonly made with iron. It consists of rotating handle used for extracting the available honey within the comb after uncapping it. This device can extract the honey from four combs simultaneously (Figure 1.2A).

- **Beeswax Extraction** is another important product of honeybees extracted for the industrial and other miscellaneous uses [1]. The comb made with beeswax is used for constructing the nest of the bees that is why after removing the maximum possible amount of honey, combs are melted either with the help of passing warm water on beeswax or using solar wax melter, which can also be used for wax processing. Beekeeper should be aware about value and importance of beeswaxes because it is used for increasing the durability of leather, water proofing, preparing threads and candle manufacturing. During extraction process, furniture polishing, shoe polishing, etc. [22, 24] should follow the suitable method for collecting maximum quantity of wax without any damage.

- **Solar Wax Melter** can be designed with the help of locally available materials like wood, plastic, glass and metal, etc. It consists of spouted shelf constructed with metal within the wooden box and entire box is covered from the top. Entire set up is exposed to the sun for increasing the internal temperature of the box for melting the combs and wax can be collected in the container placed in spouted shelf [6]. If any remains amount of honey is present, then is settled towards bottom and can easily be separated from wax (Figure 1.2B).

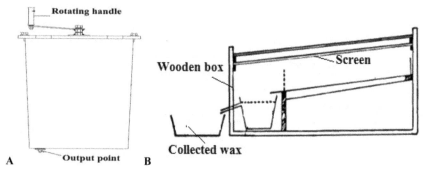

FIGURE 1.2 Equipments used for collecting honeybee's products: (A) honey extractor, and (B) solar wax melter.

1.6 PRODUCTS OF HONEYBEE AND ITS DIFFERENT USES

1.6.1 DIFFERENT PRODUCTS OF HONEYBEE

1.6.1.1 Strained Honey

Cap of wax cover of honeycomb is removed with the help of any scrapping equipment, which is placed in warm water. For collecting, the stored honey is placed in the container by cutting the frame from bottom one and finally the frame in placed into extractor after removing the wax cover. During cutting of the frame, upper portion of the top-bar is guided with one hand and other portion is situated on the provided container (Figure 1.3). Strained honey is stored in the container or in any pot after extracting and is screened by using clean cotton cloth for avoiding the any impurities in the form of wax or pollen, etc. [2]. Honeycomb is sealed with wax for protecting the honey from outer impurities, is removed with the help of scraping device due to which honey starts to come out and passes through the series of clothes screen for separating the impurities and larger particles and then the impurity free honey is collected into the container [6]. Natural honey is used for industrial, pharmaceutical and nutritional purpose thus giving good price to the producer in national and international markets [11]. Honeycomb

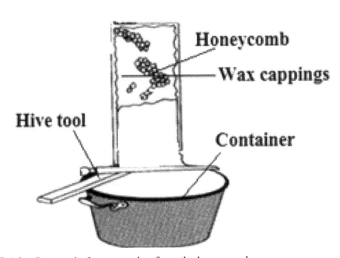

FIGURE 1.3 Removal of wax capping from the honeycomb.

also produces good quality of wax [9] and equipment like gas chromatography is used for evaluating the purity of wax components [27]. Honey is used as an important source of food during unavailability of other traditional food materials as well as during un-appropriate weather for cultivation of other main food produces [26]. Common natural honey is a viscous solution which carries 15–17% of moisture, 0.1–0.4% of protein, 80–85% of carbohydrate and 0.2% ash and other significant quantities of other substance [34, 37]. The pH of honey varies from 3.2 to 4.5 indicating the acidic characteristics [25, 28, 45, 54].

1.6.1.2   Bee Pollen

Bee pollen is extracted with help of a pollen collecting device, which is setup just before the beehives and the hind legs are scrapped and during movement of honeybee pollens are collected through holes. Bee pollens are used for making several products like cosmetics and pharmaceutical, etc. [20]. Seeley [46] stated that pollens are carried by honeybees and are deposited during return towards the hives and are also used as source of fat for the bees. Different colors of pollen can be obtained like yellow, orange, red, brown and blue.

1.6.1.3   Bee Milk

Bee milk is obtained from honeybees that contain different acids (like fatty acid and uronic acid, etc.), proteins, vitamins, phenols and reducing sugar, etc. It is beneficial for the human health like helpful to minimize the mental tension, controls the blood pressure and reduces the heavy body weight [42].

1.6.1.4   Bee Venom

Bee venom is a product of honey, and is also called as Apitoxin. It contains significant beneficial elements for health, for example, protein, amines and enzymes, etc. Generally bees release the bee venom through their stringers. A honeybee can inject 0.1 mg of venom via its stinger, which causes local inflammation and acts as an anticoagulant. Bee venom is commonly used by different pharmaceutical companies due to its medicinal and immunogenic properties for preparation of medicine for different diseases, for example, nerve pain (neuralgia), multiple sclerosis (MS), muscles problems and rheumatoid arthritis; and strengthening of immune system [44].

1.6.2   DIFFERENT USES OF HONEYBEE PRODUCTS

Honey and different other products obtained by honeybees are widely used in pharmaceuticals, meat, candle, jam and jelly, beverages and flavor industries. Honey is a most common out of all the products obtained from honeybees and is used for consumption with bread, bakeries, and cakes, etc. Honey is also used for preparation of antiseptic and skin care lotions and creams.

1.7   PROCESSING EQUIPMENT

After extraction of honey from Beehives, several processing steps (Figure 1.4) are used for retaining the qualitative characteristics and hygienic value up to the marketing. For step-wise efficient processing, different advanced equipments are needed.

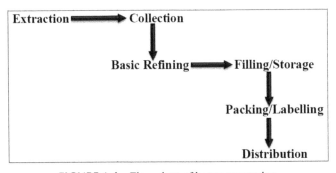

FIGURE 1.4   Flow chart of honey processing.

- *Filter Press* is used for removing the primary impurities from the collected fresh honey. Extracted amount of honey is passed through a filter, which allows the pure honey to pass through and impurities are collected in the filter.
- *Storage/Settling Tank* is used for storing the honey. When some impurities of beeswaxes also entered in it, then all beeswaxes are separated because of honey has the tendency to settle down.
- *Pre-heating Tank* is used for heating the honey for a certain period of time.
- *Processing Tank* is used for further processing of honey after collection. Maintaining of natural flavor and taste of honey is very difficult after extraction. Therefore, to get good market price, processing takes place by applying collected honey into a processing tank.
- *Cooling Tank* is used for decreasing the temperature which rises during pre-heating and collecting the pure form of honey.
- *Packing and Labeling Machine:* Different types of packaging machines are used for packing the processed honey into different sizes of packages with proper sealing. After packaging, appropriate labels is put on the packages for providing information regarding brand and other useful information.

1.8   BEEKEEPING AND HONEY PROCESSING: EMERGING ENTREPRENEURSHIP FOR RURAL AREAS

1.8.1   BEEKEEPING: AS FAMILY ACTIVITY FOR RURAL AREAS ENTREPRENEURS

Beekeeping or Apiculture is an emerging sector of agriculture for earning a source of livelihood for rural, urban and entrepreneurs communities with least investment. In Apiculture, wooden rectangular boxes contain artificial beehives that are fixed on wooden or iron stands for rearing of honeybees and due to very less space required it has become very popular source of earning among small land holders, land-less community (doing apiary on trees, building's roof and near kitchen gardens, etc.) and entrepreneurs. Apiculture is very impactful and profitable because it does not require different parameters like cropping and agricultural activities, for example, no requirement of fertile land, water resources, irrigation, fertilizers, etc. Bee boxes can be handled by one or two persons only (intensive labor involvements are not required) and least dependent on weather conditions (can be done through-out the year). It is also very helpful for the growth of crops, flowers and trees because honeybees are performing pollination for nectar collection.

1.8.2 BEEKEEPING IN INDIA

Beekeeping (also known as bee rearing or Apiculture) started widely in India since 19th century when a European origin domestic honeybee (*A. mellifera*) was introduced in Indian climate for accelerating the collection of honey and other products like wax, etc. But in beginning of 20th century, artificial wooden boxes with beehives were designed, which brought the revolution in apiculture because this box was suitable for *A. mellifera* (European species of domestic honeybees) as well as *Apix cerena Indica* (indigenous domestic species of India). Initially, boxes of honeybees were installed in Kanyakumari region of southern India. For bringing the change in the life of landless, marginal farmers and farming oriented people, many legends of India have done significant work to promote the new approach of beekeeping (set up of beehives within wooden box) like in Odisha region (Smt. Rama Devi and Sri Manmohan Chaudhary); and in Jammu and Kashmir (Sri Rajdan), etc. Besides this, due to effort of Mahatma Gandhi, beekeeping has been also considered under Rural Development Programs.

1.8.3 BENEFITS OF BEEKEEPING FOR FARMERS AND RURAL PEOPLES

It is an agriculture allied practice that can accelerate the economy and employment of local and farming communities. Apiculture (beekeeping) provides several products like honey, wax and bee-milk, etc. that are used for consumption and pharmaceutical purposes. Beekeeping provides source of livelihood, extra source of income, option of non-perishable and nutritious food to rural communities besides of fare profit with low cost inputs to the people not practicing the staple crops farming. If more people are involved in beekeeping, then income of crafts man, who framed the wooden box with hives, can also increase proportionally.

1.8.4 CHALLENGES FOR BEEKEEPER

Beekeeping is an agricultural art that requires scientific as well as practical exercises for proper management and collection of different products of honeybee. Some intensive care should be taken during beekeeping, for example, what and how much feed should be supplied to the bee-colonies during dearth period (duration in which sufficient flowers, trees and crops are not available for pollination, which affects the collection of nectars that is why some external food is required for feeding to honeybees), proper observation of timing of main nectars flow, bees population in each colony, effect of environment on bees multiplication and their activities (because in suitable environmental condition, growth of bees and honey yield both increased significantly), etc. Adulteration test of honey is also a tedious work, because of several impurities can be mixed and identification of each and every impurity is quite difficult [10].

1.9 SUMMARY

Apis mellifera is a species of honeybees widely accepted in Europe and African continent and *A. cerana*, *A. dorsata*, and *A. florae* is widely available in Asian region out of which *A. cerana* is most popular and can be of kept in beehives for honey extraction. Today beekeeping has emerged as good agriculture allied alternate of livelihood which provides another option of nutritious food, source of income and employment. Beekeeping can be performed in the kitchen garden, flower garden, crop fields and fruit garden, because no specific land is utilized.

In present scenario, honey extraction and its processing have emerged as lucrative entrepreneurship because honey is a hygienic food as well as very important for pharmaceutical (honey have anti-inflammatory and anti-oxidant properties besides it contains several amino acids and enzymes, etc.), bakery, soap, oil and other industries.

Honeybees prepare honey nectar, which is collected from different flowers, trees and crops and stored into Beehives. For proper functioning of honeybee's temperature of beehives should be maintained around 35°C.

Different products besides honey are also available like pollen, bee milk, Apitoxin and wax, etc. For several useful consumable and daily ways. Beekeeping is an eco-friendly practice and method of harnessing the naturally available and very useful honey products. Practice of beekeeping is very useful and able to enhance the yield of different crops due to pollination. In present time, it has become a good source of energy and other nutrients as well as source of income in different regions of the world and basically honey production is increasing very rapidly in developing regions like Asia and Africa. Beekeeping can be practiced in natural hives as well as manmade hives also. In modern age for accelerating the beekeeping, mostly manmade bee boxes are preferred because these are easy to access and can be supervised daily. Honey processing industries have increased significantly in 19th and 20th centuries for extraction and value addition. Today, beekeeping is included within one of the key source of livelihood of agricultural allied sector and millions of peoples of the world are based on it for their livelihood.

KEYWORDS

- Adulteration
- Adulteration test
- Agricultural business
- Agriculture
- Anticoagulant
- Apiary
- Apiculture activities
- *Apis cerana*
- *Apis cerena* indica
- *Apis dorsata*
- *Apis florea*
- *Apis mellifera*
- Apitoxin
- Artificial hives
- Bee boxes
- Bee brush
- Bee combs

- Beehives
- Bee milk
- Bee pollen
- Bee rearing
- Bee venom
- Beeswax extraction
- Beeswaxes
- Blood pressure
- Bottom board
- Combs
- Cooling tank
- Domestic honeybees
- Electrical uncapping knife
- Entrepreneurs
- Entrepreneur's communities
- Farming communities
- Fatty acid
- Filter press
- Health
- Hive stand
- Hive tool
- Honey extractor
- Honey processing
- Honey production
- Honey super
- Honey comb
- Human health
- Immune system
- Immunogenic properties
- Indigenous domestic species
- Inflammation
- Intermediate technology hive
- Labeling machine
- Management
- Mental tension
- Mesh helmet
- Movable frame hives
- Multiple sclerosis
- Nectar collection
- Nerve pain
- Neuralgia
- Packaging machine
- Pharmaceutical
- Phenols
- Pollen collector
- Pollen collector device
- Pre-heating tank
- Processing tank
- Proteins
- Reducing sugar
- Rheumatoid arthritis
- Scrapping equipment's
- Settling tank
- Smoker
- Solar wax melter
- Sting resistant gloves
- Storage tank
- Strained honey
- Strengthening
- Stringers
- Traditional apiculture
- Traditional hive
- *Trigona* spp.
- Uronic acid
- Venom
- Vitamins
- Wax capping

REFERENCES

1. Adjare, S. O., (1984). The Golden Insect: A Handbook on Beekeeping for Beginners, ITDG Publishing: Bradford, UK, pp. 112.
2. Adjare, S. O., (2016). Beekeeping in Africa. FAO Agricultural Services Bulletin 68/6, Food and Agriculture Organization of the United Nations, Rome, 1990. Accessed on 19 April URL: http://www.fao.org/docrep/t0104e/t0104e00.htm.
3. Aizen, M. A., & Harder, L. D., (2009). The global stock of domesticated honey bees is growing slower than agricultural demand for pollination. *Current Biology, 19*(11), 915–918.
4. Aizen, M. A., Garibaldi, L. A., Cunningham, S. A., & Klein, A. M., (2008). Long-term global trends in crop yield and production reveal no current pollination shortage but increasing pollinator dependency. *Current Biology, 18*(20), 1572–1575.
5. Anklam, E., (1998). Review of the analytical methods to determine the geographical and botanical origin of honey. *Food Chemistry, 63*(4), 549–562.
6. Anonymous, Honey, (2016). Processing (Practical Action Technical Brief Tools for Agriculture). Accessed on 19 April URL: http://www.appropedia.org/Honey_Processing_(Practical_Action_Technical_Brief).
7. Arias, M. C., & Sheppard, W. S., (2005). Phylogenetic relationships of honey bees (Hymenoptera:Apinae:Apini) inferred from nuclear and mitochondrial DNA sequence data. *Molecular Phylogenetics and Evolution, 37*, 25–35.
8. Atwal, A. S., & Dhaliwal, G. S., (1969). Some behavioral characteristics of Apisindica F and Apismellifera L. *Indian Bee Journal, 31*, 83–90.
9. Bogdanov, S., Imdorf, A., Kilchenmann, V., & Gerig, L., (1990). Rückstände von fluvalinat in bienenwachs, futter und honig. *SchweizerischeBienen-Zeitung, 113*(3), 130–134.
10. Bogdanov, S.,Jurendic, T.,Sieber, R.,& Gallmann, P., (2008). Honey for nutrition and health: A review. *Journal of the American College of Nutrition, 27*(6), 677–689.
11. Buba, F., Gidado, A., & Shugaba, A., (2013). Analysis of Biochemical Composition of Honey Samples from North-East Nigeria. *Biochemistry & Analytical Biochemistry, 2*(3), 2–7.
12. Burlando, B., & Comara, L., (2013). Honey dermatology and skin care. A review. *Journal of Cosmetic Dermatology, 12*(4), 306–313.
13. Calderone, N. W., (2012). Insect pollinated crops, insect pollinators and US Agriculture: *Trend Analysis of Aggregate Data for the Period 1992–2009. PLoS ONE, 7*(5), e37235.
14. Carreck, N., (2010). Honey bee colony losses - towards a greater understanding of the causes. *The Appliance of Science, 87*(1), 10–11.
15. Crittenden, A. N., (2011). The Importance of honey consumption in Human evolution. *Food and Foodways: Explorations in the History and Culture of Human Nourishment, 19*(4), 257–273.
16. Dyer, F. C., & Seeley, T. D., (1987). Interspecific comparisons of endothermy in honeybees (Apis): deviations from the expected size-related patterns. *Journal of Experimental Biology, 122*, 1–26.

17. Engel, M. S. A., (2001). monography of the Baltic amber bees and evolution of the Apoidea (Hymenoptera). *Bulletin of the American Museum of Natural History, 259*, 5–192.

18. Engel, M. S., (1999). The taxonomy of recent and fossil honey bees (Hymenoptera: Apidae: Apis). *Journal of Hymenoptera Research, 8*, 165–196.

19. Foret, S., Kucharski, R., Pellegrini, M., Feng, S., Jacobsen, S. E., & Robinson, G. E., (2002). DNA methylation dynamics, metabolic fluxes, gene splicing, and alternative phenotypes in honey bees. *Proceeding of National Academic of Science USA, 109*, 4968–4973.

20. Fujiyoshi, H., & Nakamura, J., (2009). Products of beekeeping – 8 gifts of honey bees. In: *Development of Beekeeping in developing countries and practical procedures- Case Study in Africa. Japan Association for International Collaboration of Agriculture and Forestry (JAICAF)*, Accessed on 19 April 2016,URL: http://www.jaicaf.or.jp/English/bee_en.pdf

21. Garnery, L., (1991). Phylogenetic relationships in the genus Apis inferred from mitochondrial DNA sequence data. *Apidologie, 22*, 87–92.

22. Ghosh, G. K., (1998). Beekeeping in India. Ashish Publishing House, New Delhi. India, pp. 233.

23. Hadisoesilo, S., & Otis, G., (1998). Differences in drone cappings of *Apiscerana* and *Apisnigrocincta. Journal of Apicultural Research, 37*, 11–15.

24. Hepburn, R., & Radloff, S. E., (2011). Honeybees of Asia (Hepburn, R. and Radloff, S. E. Eds.) Springer-Verlag Berlin Heidelberg, New York, pp. 669.

25. Islam, A., Khalil, I., Islam, N., Moniruzzaman, M., Mottalib, A., Sulaiman, S. A., et al., (2012). Physicochemical and antioxidant properties of Bangladeshi honeys stored for more than one year. *BMC Complementary and Alternative Medicine, 12*(177), 1–10.

26. James, O. O., Mesubi, M. A., Usman, L. A., Yeye, S. O., & Ajanaku, K. O., (2009). Physical characteristics of some honey samples from North-Central Nigeria. *International Journal of Physical Sciences, 4*(9), 464–470.

27. Jimenez, J. J., Bernal, J. L., Del-Nozal, M. J., Toribio, L., & Bernal, J., (2007). Detection of beeswax adulterations using concentration guide-values. *European Journal of Lipid Science and Technology, 109*(7), 682–690.

28. Jimoh, W. L. O., & Ummi, U. A., (2014). Chemical Characteristic of Selected Imported and Local Honey Sold Around Kano Metropolis, Nigeria. *Bayero Journal of Pure and Applied Sciences, 7*(2), 55–58.

29. Kappico, J. T., Suzuki, A., & Hongu, N., (2012). Is Honey the Same as Sugar? College of Agriculture and Life Sciences Cooperative Extension, University of Arizona. Accessed on 19 April 2016, URL: http://extension.arizona.edu/sites/extension.arizona.edu/files/pubs/az1577.pdf

30. Khakina, P. N., (2013). Modern Beekeeping: A Case Study of West Pokot Honey Processing Pilot Plant. In: *Kenya Vision 2030. Kenya Industrial Research and Development Institute (KIRDI), Nairobi.* Accessed on 19 April 2016, URL: http://www.tum.ac.ke/assets/research/sec_sti/DAY%202/MODERN%20BEEKEEPING%20A%20CASE%20STUDY%20OF%20WEST%20POKOT%20HONEY%20PROCESS-ING%20PILOT%20PLANT.pdf

31. Koeniger, N., Koeniger, G., Tingek, S., & Kelitu, A., (1906). Interspecific rearing and acceptance of queens between *Apiscerana*Fabricius, and *Apiskoschevnikovi*Buttel-Reepen. *Apidologie, 27*(5), 371–380.

32. Koetz, A. H., (2013). Ecology, behavior and control of *Apiscerana* with a focus on relevance to the Australian incursion. *Insects, 4*, 558–592.

33. Li, J. K. F. M., Zhang, L., Zhang, Z. H., Pan., & Y. H., (2008). Proteomics analysis of major royal jelly protein changes under different storage conditions. *Journal of Proteome Research, 7*, 3339–3353.

34. Miraglio, A., (2016). Honey Health and therapeutic qualities. EEUU: National Honey Board. 2001. Accessed on 19 February, URL:http://www.biologiq.nl/UserFiles/Compendium%20Honey%202002.pdf

35. Miyamoto, S., (1986). Biological studies on Japanese bees: Differences in flower relationship between a Japanese and a European honeybee. *Sci Rep Hyogo UnivAgricSerAgricBiol, 3*, 99–101.

36. Nguyen, B. K., Mignon, J., Laget, D., De-Graaf, D. C., Jacobs, F. J., Vanengelsdorp, D., rt. al., (2010). Honey bee colony losses in Belgium during the 2008-2009 winter. *Journal of Apicultural Research, 49*(4), 337–339.

37. Nichollus, J., & Miraglio A. M., (2003). Honey and healthy diets. *Cereal Food World, 48*(3), 116–119.

38. Oldroyd, B. P., & Siriwat, W., (2006). Asian Honey Bees (Biology, Conservation, and Human Interactions). Cambridge, Massachusetts and London, England: Harvard University Press, pp. 425.

39. Ono, M., Okada, I., & Sasaki, M., (1987). Heat protection by balling in the japanese honeybee *Apiscerana japonica* as a defensive behavior against the hornet, *Vespa simillimaxanthoptera* (Hymenoptera: Vespidae). *Experientia, 43*, 1031–1032.

40. Park, D., Jung, J. W., Choi, B. S., Jayakodi, M., Lee, J., & Lim, J., et al., (2015). Uncovering the novel characteristics of Asian honey bee, Apiscerana, by whole genome sequencing. *BMC Genomics, 16*(1), 1–16.

41. Peng, Y. F., Xu, S., & Ge, S., (1987). The resistance mechanism of the Asian honey Bee, *Apiscerana*fabr., to an ectoparasitic mite, varroajacobsonioudemans. *Journal of Invertebrate Pathology, 49*, 54–60.

42. Phadke, R. P., (2008). Beekeeping as an Industry and Its Role in Forestry, Agriculture and Horticulture. In: Proceedings of the Workshop on Role of Apiculture in Increasing in Crop Yields in Horticulture, Maharashtra State Horticulture and Medicinal Plants Board, Pune, India, pp. 245.

43. Pirk, C. W. W., Human, H., Crewe, R. M., & Vanengelsdorp, D., A., (2014). survey of managed honey bee colony losses in the Republic of South Africa 2009 to 2011. *Journal of Apicultural Research, 53*(1), 35–42.

44. Ramachandra, T. V., Subash-Chandran, M. D., Joshi, N. V., & Balachandran, C., (2012). Beekeeping Sustainable livelihood option in Uttarakannad, Central Western *Ghats*.Energy and wetlands research group, Centre for Ecology Sciences, Indian Institute of Science, Bangalore, India, pp. 187.

45. Saxena, S., Gautam, S., & Sharma, A., (2010). Physical, biochemical and antioxidant properties of some Indian honeys. *Food Chemistry, 118*, 391–397.

46. Seeley, T. D., The Wisdom of the Hive–The Social Physiology of Honey Bee Colonies.Harvard University Press, USA, 1996, pp. 310.

47. Seeley, T. D., & Morse, R. A., (1977). Dispersal behavior of honey Bee swarms. *Psyche, 84*(2–4), 199–209.
48. Seeley, T. D., & Seeley, R. H., (1982). Colony defense strategies of the honeybee in Thailand. *Ecological Monographs, 52, 43–63.*
49. Sivaram, V., (2012). Status, Prospects and Strategies for Development of Organic Beekeeping in the South Asian Countries. Division of Apiculture and Biodiversity, Department of Botany, Bangalore University, Bangalore, India, pp. 130.
50. Soman, A. G., & Chawda, S. S., (1996). A contribution to the biology and behaviour of dwarf bee, A. florea F. and its economic importance in Kutch, Gujarat, India: *Indian Bee Journal, 58*(2), 81–88.
51. Somerville, D., (2016). Removing Honey from the Hive. NSW Agriculture, Agnote DAI-133. 2002. Accessed on 19 April URL: http://www.dpi.nsw.gov.au/__data/assets/pdf_file/0008/117548/removing-honey-from-hive.pdf.
52. Thomas, D., Pal, N., & Subba Rao, K., (2002). Bee Management and Productivity of Indian Honeybees. *Apiacta, 3,* Accessed on 19 April 2016, URL: http://wgbis.ces.iisc.ernet.in/biodiversity/sahyadri_enews/newsletter/issue46/bibliography/81_Bee%20management%20and%20productivity%20of%20Indian%20honey%20bees.pdf.
53. White, J. W., (1957). The Composition of Honey. *Bee World, 38,* 57–66.
54. White, J. W., (1975). *Physical Characteristics of Honey.* In: Crane, E. (ed.), Honey:A Comprehensive Survey, Hienemann, London, UK, pp. 207–239.
55. Xu, P., Shi, M., & Chen, X. X., (2009). Antimicrobial peptide evolution in the Asiatic honey bee (*Apiscerana*). *Plos one, 4,* e4239.

CHAPTER 2

HERBAL FORMULATIONS FOR TREATMENT OF DENTAL DISEASES: PERSPECTIVES, POTENTIAL, AND APPLICATIONS

ASHWINI G. PATIL, PRAVIN ONKAR PATIL, ARPANA H. JOBANPUTRA, and DEEPAK KUMAR VERMA

CONTENTS

2.1 INTRODUCTION

Our Mother Nature has gifted us a variety of plants useful for medications to cure a wide range of diseases. These herbal plants are well described for medicinal uses in the history since 5000 B.C. [25]. The herbal formulations are made-up of multiple combinations of different herbs. Each of them possesses a tremendous potential for curing the dental disease is commonly used and its importance also has described in Ayurveda. The herbal formulations can nourish to our tooth by supporting fight against various organisms that cause dental diseases. The growing interest in the potential benefits of herbal formulations together with its popularity as made-up of herbs have prompted as large number of investigations for different chemical constituents and their biological activities as an economically feasible and delivered us cost-effective alternatives for curing the dental diseases [25, 43, 51].

Dental caries is an indigenous infectious microbial disease caused by cariogenic bacteria that lead to the localized dissolution and damage of the calcified tissues of the teeth. Dental caries is also known as tooth caries, cavities, or decay [37, 56–58]. They may vary in colors ranging from yellow color to black color with the general symptom of pain and eating difficulty in exposure to tooth loss, infection and finally death in severe cases. Dental caries may also be responsible for bad breath and foul tastes [36, 55, 63, 70]. In highly progressed cases, infection can spread from the tooth to surrounding soft tissues, which may lead to an edentulous mouth [22].

A number of Gram-positive bacteria are closely associated with the development and progression of dental caries [39], where the hard tissues viz. cementum, dentin and enamel of the teeth are broken down [45, 54, 60]. Organisms such as *Candida albicans, Staphylococcus aureus, Streptococcus mutans,* and *Strep. mitis* were the primary cariogenic microbes from dental plaques [1, 7, 18, 28, 39, 55]. The dental plaque refers to the sticky substance form inside the mouth with the complex combination of food debris, acid, bacteria, and saliva [33]. It has been found at the surfaces of the tooth through the production of extra cellular polysaccharides from disaccharide molecules sucrose ($C_{12}H_{22}O_{11}$) made-up of glucose and fructose monosaccharides [1, 7, 28, 39, 71]. They are subsequently broken-down into the sugar molecules to organic acid such as lactic acid which is responsible for the tooth enamel demineralization [1, 7, 28, 31]. Dental caries can be prevented by the use of important strategies such as; anti-bacterial agents employed for elimination of cariogenic bacteria from the oral cavity.

Today, dental caries remain as one of the most important and common oral health problems throughout the world [50, 55], which is associated with the people of all groups ranging from lower to higher socioeconomic status [68, 69] but generally, related with poor families where people poorly clean their mouth, and receding gums caused flouring the proper growth and development of the roots of the teeth [19, 53, 54, 56]. Dye and Thornton-Evans [20] reported that data on US children with dental caries (like filled teeth, teeth missing and untreated dental caries) indicate 24% belong to 2 years to 4 years, 53% belong to 6 years to 8 years and 56% belong to 15 year of age, respectively. Extensive efforts have been made to find an active agent against dental caries. However, an anti-cariogenic organism was found to be resistant to various anti-bacterial agents' *viz*. ampicillin, chloramphenicol, penicillin, etc. [16, 31]. Numerous studies have been concluded with respect to the herbal formulations in treatments of dental diseases. However, there is a dire need for further innovative research and progressive development on natural anti-microbial agents, which are very effective and safe for the human use.

According to the World Health Organization (WHO), it has been predicted that about 80% of the people belonging to the developing nations depend on traditionally plant-based medicines for their primary dental care. Synthetic drugs are associated with some issues for example; drug resistance, high treatment cost and side effects, etc. Traditional medicines are not only easily available but also affordable. Recently, there is increasing importance on the use of plants and plant based products to treat dental diseases. Global markets are thereby diverting to herbals that are considered as prospective and realistic source of ingredients for dental care products [12, 14].

The present chapter is an attempt towards summarizing the different herbal formulations reported in the literature for the treatment of dental diseases.

2.2 DIFFERENT HERBAL FORMULATIONS

Different herbal formulations are alternative options for prevention and treatment of dental diseases because the product formulations prepared from the herbal sources are safe, effective and economically available as compared to other choices (Table 2.1).

TABLE 2.1 Important Plants and Trees Reported for Production of Herbal Formulations in Treatment of Dental Disease

Scientific name	Common name	Family	Used plant part/extract	Reference
Acacia, arabica	Acacia Babul, Wattle Bark, Indian Gum	Legumino-sae;	Bark	[55]
Acacia senegal	Acacia	Fabaceae	Bark	[47]
Accacia nilotica or *Vachellia nilotica*	Gum arabic tree	Leguminosae (Fabaceae)	Hydro-alcoholic Extract	[27]
Achyranthes aspera	Chaff-flower, Prickly chaff flower, Devil's horsewhip	Amarantha-ceae	Hydro-alcoholic Extract	[27]
Aloe vera	Ghritkumari	Xanthorrhoe-aceae	Gel	[21]
Azadirachta indica	Neem	Meliaceae	Fruit, Bark, Hydro-alcoholic extracts, leaves, Poly-herbal extract, Neem extract	[10, 26, 27, 44, 55, 64, 65]
Berberis vulgaris	Barberry	Berberida-ceae	Alkaloid extract of roots and barks	[38]
Camellia sinensis or *Thea sinensis*	Green Tea	Theaceae	Water extract of green tea, Leaves	[40, 67]
Caryophyllus aromaticus	Clove	Myrtaceae	Bark, Fruits, Oil	[47, 64, 67]
Cinnamomum zeylanicum	Cinnamon	Lauraceae	Bark	[64, 67]
Curcuma longa	Turmeric	Zingibera-ceae	Hydro-alcoholic extracts, Rhi-zome	[27, 55]
Emblica officinalis or *Phyllanthus emblica*	Indian goose-berry (Amla)	Phyllantha-ceae	Bark, Dried pulp or Fruit, Fruit	[44, 47, 55, 64]
Ferula asafoetida	Asafetida	Fabaceae	Bark	[47]
Glycyrhhiza glabra	Liquorice	Fabaceae	Roots and Rhi-zome; Root	[55, 67]

TABLE 2.1 (Continued)

Scientific name	Common name	Family	Used plant part/extract	Reference
Mangifera indica	Mango	Anacardiaceae	Fruit, Leaves	[44, 64]
Mentha arvensis	Mentha or Mint	Lamiaceae	Leaf	[47]
Mimusops elengi	Spanish cherry, Medlar Bullet wood	Sapotaceae	Hydro-alcoholic extracts, Bark	[27, 55]
Moringa oleifera	Moringa, Drumstick tree	Moringaceae	Root	[67]
Myristica fragrans	Nutmeg	Myristicaceae	Extract of Nutmeg	[3]
Ocimum sanctum or *Ocimum tenuiflorum*	Holy basil, or tulasi (also spelled thulasi)	Lamiaceae	Leaves	[55]
Piper longum	Long pepper, Indian long pepper	Piperaceae	Fruit	[55]
Piper nigrum	Black pepper	Piperaceae	Hydro-alcoholic extracts	[27]
Psidium guajava	Guava	Myrtaceae	Aqueous extract	[13, 52]
Punica granatum	Pomegranate	Lythraceae	Phytotherapic gel	[66]
Quercus infectoria	Aleppo oak	Fagaceae	Fruit	[55, 64]
Salvadora persica or *Salvadora indica*	Arak, Galeniaasiatica, Meswak, Peelu, Pīlu, Toothbrush tree	Salvadoraceae	Hydro-alcoholic extracts, Bark	[2, 27, 55]
Sapindus mucorosai	Chinese soapberry or Washnut	Sapindaceae	Hydro-alcoholic extracts	[27]
Spilanthes acmella or *Acmella oleracea*	Toothache plant, Paracress and electric daisy.	Asteraceae	Gel	[26]
Stevia rebaudiana	Candyleaf, Sweet leaf, or Sugar leaf	Asteraceae	Leaves	[55]
Terminalia arjuna	Arjuna or Arjun tree	Combretaceae	Bark	[44]

TABLE 2.1   (Continued)

Scientific name	Common name	Family	Used plant part/extract	Reference
Terminalia belerica	Bahera, Beleric and Bastard myrobalan	Combreta-ceae	Whole fruit, Fruit	[44, 55, 64]
Terminalia chebula	Myrobalan	Combreta-ceae	Fruits and Bark	[42, 55, 64, 67]
Trachyspermum ammi	Ajowan	Apiaceae	Bark	[47]
Zanthzylum arma-tum	Prickly ash and Hercules club	Rutaceae	Hydro-alcoholic extracts	[27]
Zingiber officinale	Ginger	Zingibera-ceae	Hydro-alcoholic extracts, Bark	[27, 47]

1.2.1   HERBAL CHEWING TABLETS

Herbal chewable tablets are an important alternative and convenient to conventional tablets in dental diseases. Such tablets have great and potential merits because of no need of water. It implies that the tablets can be given to the patient at any place in any time. Chewable tablets provide additional advantages for patients to ensure better compliance, to improve the experience and to overcome swallowing difficulties [30].

The increasing prevalence of multi drug resistant strains of bacteria and the recent manifestation of strains with abridged susceptibility to antibiotics increases the prevalence of untreatable bacterial infections and poses necessity for searching new infection combating strategies in dental diseases [28]. Many popular herbal products are known to control dental plaques and gingivitis. However, they have been used for a limited period of time and used only as an adjunct to other oral hygiene measures such as brushing and flossing [35]. Chewable tablets are one of the best remedies available against cariogenic microorganisms [52].

The anti-cariogenic activity of chewable tablets of guava extract against *Strep. mutans* has been reported by Saraya et al. [52]. Rationale for the selection of guava extract was that its aqueous extracts have *in vitro* anti-bacterial effect on the growth of plaque bacteria and it may have potential use as anti-plaque agents. The guava extract chewable tablets were formulated using the classical wet granulation method [29, 52]. The tablets (1 g each) were prepared using crude extract, PVP (polyvinyl pyrrolidone) K30 (as 10% solution), mannitol, aerosil, magnesium stearate, peppermint and menthol as excipients. Results of the study revealed that 32 mg/g of the guava extract in chewable tablets exhibited best bacteriostatic activity after 15 minutes but no bactericidal activity. On the other hand, Thombre et al. [64] attempted formulation of chewable tablet using polyherbals viz. bark of *Azadirachta indica* (neem), fruit of *Caryophyllus aromaticus*, bark of *Cinnamomum zeylanicum*, fruit of *Emblica officinalis*, fruit of *Mangifera indica*, fruit of *Quercus infectoria,* fruit of *Terminalia belerica* and bark of *T. chebula.* The polyherbals chewable tablets were prepared by direct compression method using 20 mg powder of each herb per each tablet and was mixed properly with 5%w/v of PVP in alcohol as binder, 5% starch powder was added as disintegrating agent. The tablet showed good anti-bacterial activity with greater zone diameter than individual extracts tested. These studies led to the conclusion that instead of using individual extract, mixture of extracts was more effective. The tablets showed the synergistic effect on plaque formation.

1.2.2   HERBAL GELS

Herbal gels are defined as semi rigid systems in which movement of the dispersing medium is restricted by an interlacing three-dimensional network of particles from different herbs or solvated macromolecules of the dispersed phase [24]. Herbal gels have better potential as a vehicle to administer drug topically in comparison to ointment, because they are non-sticky, require low energy during formulation, are stable and have aesthetic value [41]. Herbal gels have several advantages over conventional gels in minimizing side effects and increasing the therapeutic efficacy. Herbal gels with suitable rheological properties can facilitate the absorption of poorly absorbed drug by increasing the contact time of the drug with the skin [48].

Muco-adhesive dosage forms have been used to target local disorders at the mucosal surface to reduce the overall dosage required and to minimize the side effects that may be caused by the systemic administration of the drugs. In these formulations, polymers were used as the adhesive component. These polymers are often water-soluble and when used in a dry form, they attract water from the mucosal surface and this water transfer leads to a strong interaction. These polymers also form viscous layers when hydrated with water, which increases the retention time over the mucosal surfaces and leads to adhesive interactions. Herbal gels have also been reported in literature for the treatment of various dental disorders [26, 27].

The anti-microbial potential of *Spilanthes acmella* (Akkalkara plant) gel against the different microorganisms (*Bacillus cereus, Escherichia coli, Lactobacillus* and *Streptococcus*), which are responsible for causing tooth decay, was reported by Gupta et al. [26]. The gel from Akkalkara was prepared using carbopol 934 and sodium carboxy methyl cellulose (SCMC) as a gelling agent. Carbopol 934 was dispersed in preserved water (sodium metabisulphate 0.05%) and glycerin overnight. The extract was dissolved in above solution and stirred for 10 minutes and neutralized by triethanolamine to pH 6.4 and then mixed at 300 rpm for 10 minutes [23]. Formulated gel was evaluated for spreadability, pH, homogeneity, viscosity, *in-vitro* diffusion study, muco-adhesion measurement and stability. The values of surface pH within the range of neutral or slightly acidic indicated that such formulations could be used without any irritation in the oral cavity. Spreadability of gel formulations was found to be in the acceptance range. Results of anti-microbial study revealed that 5% concentration of extract of *S. acmella* showed pronounced anti-microbial action, comparable to the standard drug moxifloxacin.

A clinical study on the pomegranate (*Punica granatum*) extract gel was studied by Somu et al. [59] against gingivitis. Fresh extract was obtained from pomegranate seeds and then it was dissolved in 5 g of CMC in 100 ml of distilled water and stirred gently for 15 minutes until the gel attained a consistency (0.05%) convenient for usage. A very small amount of methyl paraben (2 mg) was added as a preservative. The control gel was also prepared, which had the same formulation except the extract of *Pun. granatum*. The results of Somu et al. [59] were conclusive of the fact that the gel containing *Pun. granatum* extract was effective in treatment of gingivitis when used along with mechanical cleaning in controlling plaque and gingivitis.

Alternatively, Makarem et al. [38] reported efficacy of aqueous extract gel of barberry (*Berberis vulgaris*) for the control of plaque and gingivitis. A dental gel was prepared with the help of soxhlet method using alkaloid extract of roots and barks of *Ber. vulgaris* plant as the test material. The gel containing 1% berberine was formulated at pH 5, a placebo gel (the dental gel without *Ber. vulgaris* extract) was also prepared. In this study it was reported that the *Ber. vulgaris* dental gel was effective in controlling microbial plaque and gingivitis in school aged children; considering the fact that no side effects were observed with the dental gel with *Ber. vulgaris* during the study period. This gel could be recommended to be used as a dentifrice.

Fani et al. [21] reported inhibitory activity in the gel of *Aloe vera* on some clinically isolated cariogenic and periodontopathic bacteria. In this study, the inhibitory activities were observed in the gel of *Aloe vera* on some cariogenic (*Strep. mutans*), periodontopathic (*Aggregatibacter actino mycetemcomitans, Porphyromonas gingivalis*) and an opportunistic periodonto pathogen (*Bacteroides fragilis*) isolated from patients with dental caries and periodontal diseases. Salient findings of the studies were: *Strep. mutans* was most sensitive to *Aloevera* gel with a MIC of 12.5 μg/ml, while *Aggre. actinomycetemcomitans, Bact. fragilis* and *Porph. gingivalis* were less sensitive, with a MIC of 25–50 μg/ml ($p < 0.01$). Therefore, it was concluded that *Aloe vera* gel at optimum concentration could be used as an antiseptic for prevention of dental caries and periodontal diseases [21].

Vasconcelos et al. [66] conducted experiment on anti-microbial effect of a *Pun. granatum Linn.* (pomegranate) phytotherapic gel against three standard *Streptococci strains* viz. *Strep. mutans* ATCC 25175, *Strep. sanguis* ATCC 10577 and *Strep. mitis* ATCC 9811. *Strep. mutans* (clinically isolated) and *C. albicans* (either alone or associated with other microorganisms) [66]. The gel of *Pun. granatum* had better efficiency as compared to miconazole. The gel of *Pun. granatum* was also reported to be effective in inhibiting the adherence of the *C. albicans, Strep. mitis, Strep. mutans* and *Strep. sanguis*. The findings of Vasconcelos et al. [66] suggested the possibility that the gel of *Pun. granatum* might be useful in controlling bacteria and yeasts that cause various oral infections namely caries, periodontal disease and stomatitis.

2.2.3   POLYHERBAL FORMULATIONS

Polyherbal formulations refer to formulations in which two or more than two herbs are involved to develop a product [61] to reduce adverse events by enhancing the therapeutic action and reducing the single herbs concentrations [49]. Sometimes, these herbs are combined with mineral preparations [61]. The concept of polyhedral formulation is well-established and documented in the literature of ancient history [49]. Today, a challengeable task is subjected to concern on the development of a stable polyherbal formulation because of the wide diversity of chemical compounds in different medicinal plants [8]. They are compilation of therapeutic entities that are formulated and prepared on the basis of the healing properties of individual ingredients with respect to the condition of sickness. Such herbal constituents with diverse pharmacological activities principally work together in a dynamic way to produce maximum therapeutic benefits with minimum side effects [6, 9]. Different polyherbal formulations are available in today's market [32].

A polyherbal formulation was used by Sharma et al. [55] to prepare toothpaste and evaluate its anti-microbial activity. In the composition of toothpaste formulation, the plant parts of various polyherbals viz. bark of *Acacia arabica, A. indica, Mimusops elengi, Salvadora persica;* fruit of *E. officinalis,* fruit of *Piper longum, Terminalia belirica, T. chebula, Q. infectoria;* leaves of *Ocimum sanctum* or *Ocimum tenuiflorum, Stevia rebaudiana;* rhizome of *Curcuma longa;* root of *Glycyrrhiza glabra* were used. The study showed significant anti-microbial activity against all selected human oral pathogens viz. *Strep. mutans* and *Candida albicans.* The findings of Sharma et al. [55] supported and suggested that medicinal herbs and plants possess anti-microbial properties, are traditionally applied and can be employed in polyherbal formulation to prevent various dental diseases.

Pathak et al. [44] had mentioned anti-microbial activity of a polyherbal extract against dental micro flora. Total 20 g of poly-herbal powder of leaves of *A. indica,* dried pulp or fruit of *Emblica officinales,* leaves of *M. indica,* bark of *T. arjuna,* whole fruit of *T. belerica* and *T. chebula* in the proportion of 1:0.25:0.25:1.5:1:1, respectively was cold macerated using 100 mL mixture of ethanol and water (in proportion of 70:30). The filtrate was concentrated to yield 20 mL of a semisolid extract [12]. The *A. indica* was reported to be used widely in oral care formulations. *E. officinales, T. belerica* and *T. chebula,* were used because they appear to be synergistic to the anti-microbial activity of *A. indica* in maintenance of oral hygiene. *M. indica* and *T. arjuna* were selected because of their astringent and antioxidant properties in addition to anti-microbial activity. The polyherbal extract of 10% w/v concentration was found to be an effective anti-microbial formulation which was quite safe in animal toxicity studies. Therefore, formulation could also be routinely used for improving oral hygiene of healthy children and adults as well as in patients with dental caries and gingivitis [44].

The polyherbal formulation of hydro-alcoholic extracts of *A. indica, Accacia nilotica, Achyranthes aspera, Curcuma longa, Mim. elengi, Pip. nigrum* (Black pepper), *Sal. persica, Sapindus mucorosai, Syzygium aromaticum, Zanthzylum armatum* and *Zingiber officinale* have also been tried. The observations of this study clearly suggest the suitability of polyherbal formulation as herbal remedy for maintaining oral hygiene, since it possesses potent anti-microbial activity against bacterial (*B. subtilis, E. coli, Micrococcus luteus, Pseudomonas aeruginosa, Staph. aureus, Strep. mutans*) and yeast (*C. albicans*) strains [27].

2.2.4 HERBAL LOLLYPOP

Herbal lollipops are sugar-free products made up of licorice extracts using a standard sugar-free candy formula suggested by Dr. John's Candy [11]. The main ingredients of formulations were: hydrogenated starch hydrolysate (HSH) (solidifying agent), citric acid and mint (flavoring agents), FD & C blue #1,2; Red 3, 40; Yellow 5, 6 (coloring agents), and acesulfame potassium (non-caloric sweetener). Licorice extracts (about 7–15 mg) were added to each lollipop to make the uniformity of Glycyrrhizol A concentration. The amounts of licorice extracts are always dependent upon the Glycyrrhizol A concentration in a batch. The lollipop manufacturing comprises various temperatures ranges from maximum 135°C to minimum 65°C for cooking syrup and on the cooling table, respectively. At various temperatures, the thermal stability of Glycyrrhizol A was tested.

Then, the extracts of the herbal are added at the particular and appropriate temperatures at which their bioactivities could not be affected [15]. The studies showed that the sugar-free lollipops were safe and their anti-microbial activities against cavity causing bacteria were stable in the formulations intended for delivery. It was also revealed from pilot human studies that a brief application of these lollipops (twice a day for 10 days) led to marked reduction of cavity-causing bacteria in oral cavity among most of the human subjects under study. Therefore, lollipops turn out to be a useful delivery system well accepted by different populations with problems in oral cavity [15].

2.2.5   HERBAL MOUTHWASHES

Mouthwashes are concentrated aqueous solutions of anti-microbial preparations, routinely used in the oral cavity after dilution to counter infections, for cleansing and anti-sepsis as well as refreshing the oral cavity. The health benefits of herbal mouthwashes include: relieving symptoms of gingivitis, canker sores, inflamed gums, sore mouth, inflamed or ulcerated throat, mouth infections, bleeding gums and teeth sensitivity. Mouthwashes are commonly for oral hygiene and in the delivery of active agents to the teeth and gums. The potency of these rinses to influence the plaque formation and to alter the course of gingival inflammation has been reviewed extensively by many researchers. A decoction of the root-bark was suggested as a mouthwash for swollen gums and decoction of the leaves makes an efficacious gargle for swollen gum and ulceration of the mouth and also for bleeding gums.

A mouthwash was prepared from alcoholic extraction of *Sal. persica* chewing sticks (10 mg/ml) and was tested against dental plaque formation. The 200 g of *Sal. persica* chewing sticks were cut using a sharp knife, then grounded to powder using a food blender. The powder was extracted with 60% ethanol; the mixture was left for 24 hours, then filtered through Whatman No.1 filter paper and was autoclaved at 40°C until it was dry. The *in-vitro* activity of *Sal. persica*, revealed that its anti-bacterial action was concentration dependent; where 10 mg/ml solution produced the greatest zone of inhibition around each paper disc in the agar diffusion assay [2]. Literature revealed that the herbal mouthwash can be effectively used as an alternative to chlorhexidine and other synthetic drug containing mouthwash owing to their reported side effects [4, 5].

The anti-bacterial effect of Neem (*A. indica*) mouthwash was against salivary levels of *Strep. mutans* and *Lactobacillus,* when tested along with its effect in reversing incipient carious lesions [65]. The growth of *Strep. mutans* was inhibited by neem mouthwashes. In some other study reported on the same plant, the purpose was to compare the short-term efficacy and safety of the *A. indica* mouth rinse on gingival inflammation and microbial plaque. A double masked, randomized, parallel armed study was carried out to assess the efficacy of an oral mouth rinse based on leaves of the neem tree to reduce gingivitis [10]. The ethanolic extract of neem was used for preparation of mouthwash. Final formulation was achieved using 25% of *A. indica extract,* 20.0% of saccharine, peppermint oil (<0.1%) as flavor and amaranth red color. Results of the study revealed that a mouth rinse based on the neem tree was equally effective in reducing periodontal indices, gingival plaque and bleeding indexes. Additionally, the count of cariogenic bacteria in the saliva was reduced drastically. Moghbel et al. [40] reported a nontoxic, safe and stable mouthwash formulation of water extract of green tea (*Camellia sinensis*). It had high tannin content with antioxidants and anti-microbial potential.

A mouth rinse containing propolis was evaluated against dental plaque accumulation [34]. Results of this study revealed that mouth rinse containing propolis was thus efficient in reducing supragingival plaque formation and plaque index formation under conditions of high plaque accumulation.

Al-Saffar et al. [3] reported anti-plaque activity of mouthwash of nutmeg (*Myristica fragrans*) extract. Results showed significant anti-plaque and anti-inflammatory effects (tested by reduction in bleeding index) in comparison with the conventional chlorhexidine mouth rinse which encouraged its use in the treatment of gingival inflammation because it was a natural plant devoid from any chemical agent.

Effectiveness of mouthwash formulated from fruits of *T. chebula* was evaluated on a salivary *Strep. mutans* [42]. Mouthwash was formulated by adding 2.5% of ethanolic extract to distilled water, sodium CMC was added to provide viscosity, mannitol was added to mask the astringent effect of extract and methyl paraben was used as preservative. It was observed that there was 44.42% reduction of salivary *Strep. mutans* colony forming units in 5 minutes after rinsing.

Charles et al. [13] formulated *Psidium guajava* mouthwash and screened for anti-microbial activity against cultures of *C. albicans, E. coli* and *Staph. aureus*. Mouthwash was prepared using aqueous extract of *Ps. guajava* with sodium lauryl sulphate, peppermint emulsion; double strength chloroform water. The formulation with aqueous extract of *Ps. guajava* showed the highest anti-microbial activities due to presence of bioactive compounds in extract. Potencies of some formulations were enhanced by the addition of sodium lauryl sulphate and its absence responsible for the reduced potency irrespective of the strength of the chloroform water used as preservative. The *Ps. guajava* mouthwashes were effective at all tested concentrations against of *Staph. aureus* than *E. coli*.

2.2.6 HERBAL TOOTHPASTE AND TOOTHPOWDER

Herbal toothpaste consists of a formulation of well-constituted herbs that ensure anti-bacterial and gum tightening properties and provide absolute dental care. It contains natural taste of several ingredients like elaichi (*Elettaria cardamomum*), lavang (*Syzygium aromaticum,* also known as cloves), etc. and helps users in maintaining a fresh mouth for the whole day and also provides ideal protection against dental issues like gum bleeding and sensitivity. Further, the toothpaste is made of uncommon herbs that are safe to use and have great effect on oral hygiene and health.

The herbal toothpaste is based on ancient and well documented Ayurvedic medical formulation and gives high protection against cavities in addition to other dental and gum related issues. Herbal toothpaste not only protects from germs producing dental plaque, but it also has anti-oxidant properties owing to mixture of extracts which could not be found in conventional toothpaste [46, 62]. A tooth paste containing *Gymnema sylvestre* was evaluated against *C. albicans, Strep. aureus, Strep. mitis* and *Strep. mutans* [17]. Toothpaste was formulated using 1.5 grams of gum tragacanth in water in one container. The 56 grams of calcium carbonate, 2%w/w of *G. sylvestre* hydro alcoholic extract, 1.0 g of sodium lauryl sulphate were further added in another container and dry mixed. The 22 g of glycerin was added and mixed well until the mass became slightly wet. Then gum tragacanth was added to it and wetted completely followed by thorough mixing. The masses clumps were mixed well with water followed by addition of 0.1 g of saccharine sodium and preservative like sodium benzoate in sufficient quantity and mixing to get thick paste. Finally the peppermint oil was added in sufficient quantity as a flavoring agent. Devi et al. [17] had also formulated tooth powder containing *G. sylvestre*. It was formulated using 92.8 g of calcium carbonate and *G. sylvestre* (2%w/w) and mixed thoroughly. To the above mixed dry powder 6.0 g of sodium lauryl sulphate was added and mixed evenly. Around 0.2 g of powdered saccharine sodium, peppermint flavor was added in sufficient quantity and mixed completely and packed in well closed tight container.

Herbal tooth powder is a tooth-cleaning agent that is almost entirely made from all-natural ingredients to refresh breath, help heal gums, rid teeth of bacteria and reduce the amount of inflammation in the mouth. Herbal tooth powder has been around for centuries. The constituents of tooth powder and tooth pastes were same except that tooth powders do not contain humectants, water and binding agents. The primary function of tooth powder was the cleaning of the accessible surfaces of the teeth. Vohra et al. [67] reported efficacy of a herbal toothpowder consisting of bark of *C. zeylanicum*, roots and rhizome of *Glycyrhhiza glabra*, root of *Moringa oleifera* (Moringa or Drumstick tree), oil of *S. aromaticum* (clove), fruit of *T. chebula* and leaves of *Thea sinensis* against microbial flora of oral cavity and its comparison with marketed toothpastes [67]. The root and rhizome of *Gly. glabra*, root of *Mor. leifera*, *T. chebula* were finely powdered into uniform size taking care that no separation occurred on shaking. Clove oil (1 g) was added to the powder followed by proper mixing. The formulation was stored in an airtight wide mouth container for future use. The constituents present in toothpowder were well known for their activity against the microorganisms found in oral cavity. They have been found effective against the dental pathogens such as *Lactobacillus acidophilus*, *Streptococci salivarius*, *Streptococci sanguis*, *Strep. aureus* and *Strep. mutans*. The herbal tooth powder showed a reduction in oral bacterial count which may be due to the presence of active ingredients, natural extracts and blends of natural oil ingredients which may have anti-bacterial effects. Pawar et al. [47] prepared herbal toothpowder using bark of acacia (*Acacia senegal*), ajowan (*Trachyspermum ammi*), alum, amla (*E. officinalis or Phyllanthus emblica*), asafetida (*Ferula asafoetida*), clove (*C. aromaticus*), ginger, leaf of menthe (*Mentha arvensis*), mustard oil, neem bark and pepper and evaluated for anti-microbial activity against *C. albicans, E. coli, P. aeruginosa* and *Staph. aureus* [47].

2.3   CONCLUSIONS

Increase in the prevalence of resistance of an anti-cariogenic organism to many of the anti-bacterial agents rationalizes the importance of natural anti-microbial agents that are effective and safe for the human beings. Amongst all reviewed herbal formulations, herbal gel and herbal mouthwashes are found to be most widely accepted for the treatment of various dental diseases.

2.4   SUMMARY

Oral infections are most common among the other frequently occurring diseases worldwide which ultimately lead to dental caries and periodontal diseases. Dental treatments are highly expensive remedial measures nowadays and they incorporate the involvement of antiseptics as well as anti-bacterial agents and many more. There is an increasing prevalence of multi drug resistant strains of bacteria. The recent manifestation of strains has ended up with abridged susceptibility to antibiotics, consequently raising the number of untreatable bacterial infections and intense need to explore new infection combating strategies in dental diseases. Several anti-bacterial compounds have been isolated from a large number of plant species throughout the world. They have acquired unique position for the treatment of oral diseases. Authors have summarized the different herbal formulations for the treatment of dental diseases.

KEYWORDS

- *Acacia arabica*
- *Acacia indica*
- *Acacia senegal*
- *Accacia nilotica*
- Acesulfame potassium
- *Achyranthes aspera*
- *Acmella oleracea*
- *Aggregatibacter actino mycetemcomitans*
- Akkalkara
- Akkalkara plant
- *Aloe vera*
- Ampicillin
- Anedentulous mouth
- Anti-bacterial
- Anti-bacterial agents
- Anti-cariogenic
- Anti-cariogenic activity
- Anti-cariogenic organism
- Anti-inflammatory effects
- Anti-microbial action
- Antioxidant properties
- Anti-plaque agents
- Astringent
- *Azadirachta indica*
- *Bacillus cereus*
- Bacteria
- *Bacteroides fragilis*
- *Berberis vulgaris*
- Biological activities
- Bleeding gums
- Brushing
- Calcified tissues
- *Camellia sinensis*
- Candida albicans
- Canker sores
- Caries
- Cariogenic
- Cariogenic bacteria
- Cariogenic microbes
- Cariogenic microorganisms
- *Caryophyllus aromaticus*
- Cavities
- Cavity-causing bacteria
- Chemical constituents
- Chewable tablets
- Chewing sticks
- Chloramphenicol
- *Cinnamomum zeylanicum*
- Coloring agents
- Conventional tablets
- *Curcuma longa*
- Decay
- Demineralization
- Dental
- Dental caries
- Dental disease
- Dental disorders
- Dental micro flora
- Dental plaque accumulation
- Dental plaque formation
- Dental plaques
- Disaccharide
- Diseases
- Drug moxifloxacin
- Drug resistance
- Drug resistant strains
- *Emblica officinalis*

- Enamel
- *Escherichia coli*
- Extra cellular polysaccharides
- *Ferula asafetida*
- Filled teeth
- Flavoring agents
- Flossing
- Food blender
- Food debris
- Fructose
- Gel formulations
- Gelling agent
- Gingival inflammation
- Gingivitis
- Glucose
- *Glycyrhhiza glabra*
- Glycyrrhizol A
- Gram-positive bacteria
- *Gymnema sylvestre*
- Hard tissues
- Healing properties
- Health
- Healthcare
- Healthcare products
- Herbal
- Herbal chewing tablets
- Herbal formulations
- Herbal gels
- Herbal lollipop
- Herbal mouthwashes
- Herbal plants
- Herbal sources
- Herbal tooth powder
- Herbal toothpaste
- Human diseases
- Human oral pathogens
- Hydro-alcoholic extracts
- Hydrogenated starch hydrolysate
- Indigenous
- Infectious
- Inflamed gums
- Lactic acid
- *Lactobacillus*
- *Lactobacillus acidophilus,*
- Licorice extracts
- *Mangifera indica*
- Medicinal uses
- *Mentha arvensis*
- Microbes
- Microbial
- Microbial disease
- Microbial plaque
- *Micrococcus luteus*
- Microorganisms
- *Mimusops elengi*
- Molecules
- Monosaccharides
- *Moringa oleifera*
- Mouth infections
- Mouthwashes
- Muco-adhesion
- Muco-adhesive dosage
- Myristica fragrans
- Natural antimicrobial agents
- Neem
- Non-caloric sweetener
- *Ocimum sanctum*
- *Ocimum tenuiflorum*
- Oral cavity
- Oral health
- Oral health problems
- Oral hygiene
- Oral hygiene measures
- Oral infections
- Organic acid
- Organism

REFERENCES

1. Akhtar, M. S., & Bhakuni, V., (2004). Streptococcus pneumoniaehyaluronatlyase: an overview. *Current Science, 86,* 285–295.
2. Al-Bayaty, F. H., AI-Koubaisi, A. H., Wahid, A. N. A., & Abdulla, M. A., (2004). Effect of mouth wash extracted from *Salvadorapersica*(Miswak) on dental plaque formation, A clinical trail. *Journal of Medicinal Plants Research, 4*(14), 1446–1454.
3. Al-Saffar, M. T., Al-Talib, R. A., Taqa, G. A., & Taqa, A. A., (2008). The Effect of New Formula (Nut Meg Extract) As A Mouth Wash Compared With Chlorhexidine Mouth Wash. *Al–Rafidain Dental Journal, 8*(2), 189–196.
4. Asiri, F. Y., Alomri, O. M., Alghmlas, A. S., Gufran, K., Sheehan, S. A., & Shah, A. H., (2016). Evaluation of efficacy of a commercially available herbal mouthwash on dental plaque and gingivitis: A double-blinded, parallel, randomized, controlled trial. *Journal of International Oral Health, 8*(2), 224–226.
5. Bagchi, S., Saha, S., Jagannath, G. V., Reddy, V. K., & Sinha, P., (2015). Evaluation of efficacy of a commercially available herbal mouthwash on dental plaque and gingivitis: A double-blinded parallel randomized controlled trial. *Journal of Indian Association of Public Health Dentistry, 13*(3), 222–227.
6. Barik, C. S., Kanungo, S. K., Tripathy, N. K., Panda, J. R., & Padhi, M. A., (2015). Review on therapeutic potential of polyherbal formulations. *International Journal of Pharmaceutical Science and Drug Research, 7*(3), 211–228.
7. Bhattacharya, S., Virani, S., Zavro, M., & Hass, G. J., (2003). Inhibition of Streptococcus mutans and other oral Streptococci by (*Humuluslupulus* L.) constituents. *Economic Botany, 57,* 118–125.
8. Bhope, S. G., Nagore, D. H., Kuber, V. V., Gupta, P. K., & Patil, M. J., (2011). Design and development of a stable polyherbal formulation based on the results of compatibility studies. *Pharmacognosy Research, 3*(2), 122–129.
9. Bhusari, V. S., & Bhusari, G, S., (2012). Effect of madhumeh an antibiotic polyherbal formulation on carbohydrate metabolism and antioxidant defense in Streptozotocin-Nicotinamide (STZ-NICO) induced diabetic rats. *International Journal of Pharmacy & Life Sciences, 3*(5), 1690–1695.
10. Botelho, M. A., Santos, R. A., Martins, J. G., Carvalho, C. O., Paz, M. C., & Azenha, C., et al., (2008). Efficacy of a mouth rinse based on leaves of the neem tree (*Azadirachtaindica*) in the treatment of patients with chronic gingivitis: A double-blind, randomized, controlled trial. *Journal of Medicinal Plants Research, 2*(11), 341–346.
11. Candy, J., (2016). Dr. John's Thoughtfully Crafted Sweets. Accessed on 18 May, URL: http://www.drjohns.com/
12. Chakraborty, A., Devi, R. K. B., Rita, S., Sharatchandra, K., & Singh, T. I., (2004). Preliminary studies on anti-inflammatory and analgesic activities of *Spilanthesacmellain* experimental animal models. *Indian Journal of Pharmacology, 36*(3), 148–150.
13. Charles, O. E., Chukwuemeka, S. N., Ubong, S. E., Ifeanyichukwu, R. I., & Chidimma, S. O. (2007). A case of for the use of herbal extracts in oral hygiene: The Efficacy of *Psidiumguajava* based mouthwash formulations. *Research Journal of Applieed Sciences, 2*(11), 1143–1147.

14. Chopra, R. N., Nayara, S. L., & Chopra, I. C., (1956). Glossary of Indian medicinal plants. Council of Scientific & Industrial Research, New Delhi, pp. 168–169.

15. Chu-hong, H., Jian, H., Randal, E., Xiao-yang, W., Li-na, L., & Yan, T., et al., (2011). Development and evaluation of a safe and effective sugar-free herbal lollipop that kills cavity-causing bacteria. *International Journal of Oral Science, 3,* 13–20.

16. Crig, A., (1998). Antimicrobial resistance, danger signs all around. *Tennessee Medicine, 91,* 433–455.

17. Devi, B. P., & Ramasubramaniaraja, R., (2010). Pharmacognostical and antimicrobial screening of *Gymnemasylvestre*r.br, and evaluation of gurmar herbal tooth paste and powder, composed of Gymnemasylvestrer.br, extracts in dental caries. *International Journal of Pharma and Bio Sciences, 1*(3), 1–16.

18. Dwivedi, D., Kushwah, T., Kushwah, M., & Singh, V., (2011). Antibiotic susceptibility pattern against pathogenic bacteria causing Dental Caries. *South Asian Journal of Experimental Biology, 1,* 32–35.

19. Dye, B. A., & Li, X., (2012). Beltran-Aguilar ED. Selected oral health indicators in the United States, 2005-2008. *NCHS Data Brief, 96,* 1–8.

20. Dye, B. A., & Thornton-Evans, G., (2010). Trends in oral health by poverty status as measured by Healthy People 2010 objectives. *Public Health Reports, 125*(6), 817–830.

21. Fani, M., & Kohanteb, J., (2012). Inhibitory activity of Aloe vera gel on some clinically isolated cariogenic and periodontopathic bacteria. *Journal of Oral Science, 54*(1), 15–21.

22. Featherstone, J. D., (2000). The science and practice of caries prevention. *American Dental Association, 131,* 887–889.

23. Gandhi, R. B., & Robinson, J. R., (1994). Oral cavity as a site for bioadhesive drug delivery. *Advanced Drug Delivery Reviews, 13*(1–2), 43–74.

24. Goldman, P., (2001). Herbal medicines today and the roots of modern pharmacology. *Annals of Internal Medicine, 135*(8), 594–600.

25. Groppo, F. C., Bergamaschi, C. C., Kogo, K., Franz-Montan, M., Motta, R. H. L., & Andrade, E. D., (2008). Use of Phytotherapy in Dentistry. *Phytotherapy Research, 22,* 993–998.

26. Gupta, N., Patel, A. R., & Ravindra, R. P., (2012). Design of akkalkara (*Spilanthesacmella*) formulations for antimicrobial and topical anti-inflammatory activities. *International Journal of Pharma and Bio Sciences, 3*(4), 161–170.

27. Gupta, P., Gupta, V. K., Tewari, N., Pal, A., Shanker, K., & Agrawal, S., et al., (2012). A poly-herbal formulation from traditionally used medicinal plants as a remedy for oral hygiene. *African Journal of Pharmacy and Pharmacology, 6*(46), 3221–3229.

28. Hamada, S., & Slade, H. D., (1980). Biology, immunology, and cariogenicity of *Streptococcus mutans*. *Microbiological Reviews, 44,* 331–384.

29. Herbert, L., Lachman, L., & Schwartz, B. J., (1990). (eds.).Pharmaceutical dosage forms: Tablets, volume 3. 2ndEd, Mercel Dekker Inc, pp. 550.

30. Hermes-Pharma., (2016). Chewable Tablets. Hermes Pharma, Germany. Accessed on 28 April URL: http://www.hermes-pharma.com/our-dosage-forms/chewable-tablets.html.

31. Jarvinen, H., Jenevou, J., & Huovinen, P., (1993). Susceptibility of Streptococcus mutansto chlorhexidine and six other antimicrobial agents. *Antimicrob. Agentschemother, 37*, 1158–1159.

32. Kaur, M., & Valecha, B., (2014). Diabetes and Antidiabetic Herbal Formulations: An Alternative to Allopathy. *European Journal of Medicine, 6*(4), 226–240.

33. Kleinberg, I. A., (2002). Mixed-bacteria ecological approach to understanding the role of the oral bacteria in dental caries causation: an alternative to *Streptococcus mutans* and the specific-plaque hypothesis. *Critical Reviews in Oral Biology and Medicine, 13*, 108–125.

34. Koo, H., Cury, J. A., Rosalen, P. L., Ambrosano, G. M., Ikegaki, M., & Park, Y. K., (2002). Effect of a mouth rinse containing selected propolis on 3-day dental plaque accumulation and polysaccharide formation. *Caries Research, 36*, 445–448.

35. Kukreja, B. J., & Dodwad, V., (2012). Herbal Mouthwashes A Gift of Nature. *International Journal of Pharma and Bio Sciences, 3*(2), 46–52.

36. Laudenbach, J. M., & Simon, Z., (2014). Common Dental and Periodontal Diseases: Evaluation and Management. *The Medical Clinics of North America, 98*(6), 1239–1260.

37. Loesche, W. J., (1986). Role of *Streptococcus* mutans in human dental decay. *Microbiological Reviews, 50*, 53–380.

38. Makarem, A., Khalili, N., & Asodeh, R., (2007). Efficacy of barberry aqueous extracts dental gel on control of plaque and gingivitis. *Acta MedicaIranica, 44*(2), 91–94.

39. Marsh, P., & Martin, M., (1992). *Oral Microbiology.* 3rd ed., Chapman and Hall: London, pp. 133–166.

40. Moghbel, A., Farajzadeh, A., Aghel, N., Agheli, H., & Raisi, N., (2009). Formulation and evaluation of green tea mouthwash: A new, safe and nontoxic product for children and pregnant women. *Toxicology Letters, 189*(1), S257-S273.

41. Mohammed Haneefa, K. P., Mohanta, G. P., & Nayar, C., (2012). Formulation and Evaluation of Herbal Gel of Basellaalba for wound healing activity. *Journal of Pharmacological Science and Research, 4*(1), 1642–1648.

42. Nayak, S. S., Ankola, A. V., Mdgud, S. L., & Bolmal, V., (2012). Effectiveness of mouth rinse formulated from ethanolic extract of Terminalia chebula fruits on salivary streptococcus mutans among 12 to 15 year old school children of Belgaum city: A randomized field trial. *Journal of Indian Society of Pedodontics and Preventive Dentistry, 30*(3), 231–236.

43. Oliveira, F. Q., Gobira, B., Guimarães, C., Batista, J., Barreto, M., & Souza, M., (2007). Plantspeciesindicated in odontology. *Revista Brasileira de Farmacognosia, 17*, 466–476.

44. Pathak, A. A., Patankar, R. D., Galgatte, U. C., Paranjape, S. Y., Deshpande, A. S., & Pande, A. K., et al., (2011). Antimicrobial activity of a poly-herbal extract against dental micro flora. *Research Journal of Pharmaceutical, Biological and Chemical Sciences, 2*(2), 533–539.

45. Pathak, A., Sardar, A., Kadam, V., Rekadwad, B., & Karuppayil, S. N., (2012). Efficacy of some medicinal plants against human dental pathogens. *Indian Journal of Natural Products and Resources, 3*(1), 123–127.

46. Patki, P. S., Mazumdar, M., Majumdar, S., & Chatterjee, A., (2013). Evaluation of the safety and efficacy of complete care herbal toothpaste in controlling dental plaque, gingival bleeding and periodontal diseases.*Journal of Homeopathy & Ayurvedic Medicine*. *2*, 124.

47. Pawar, C. R., Gaikwad, A. A., & Kadtan, R. B., (2011). Preparation and evaluation of herbal tooth powder composed of herbal drugs with antimicrobial screening. *Indo American Journal of Pharmaceutical Research*, *3*, 196–202.

48. Pawar, D. P., & Shamkuwar, P. B., (2013). Formulation and evaluation of herbal gel containing lantana camara leaves extract. *Asian Journal of Pharmaceutical and Clinical Research, 6*(3), 122–124.

49. Petchi, R. R., Vijaya, C., & Parasuraman, S., (2014). Antidiabetic activity of poly-herbal formulation in streptozotocin nicotinamide induced diabetic wistar rats. *Journal of Traditional and Complementary Medicine, 4*(2), 108–117.

50. Petersen, P. E., (2003). The World Oral Health Report 2003: Continuous Improvement of Oral Health in the 21st Century: The Approach of the WHO Global Oral Health Programme. *Community Dentistry and Oral Epidemiology, 31*(Suppl. 1), 3–23.

51. Pieri, F. A., Mussi, M. C. M., Fiorini, J. E., Moreira, M. A. S., & Schneedorf, J. M., (2012). Bacteriostatic effect of copaiba oil (*Copaifera officinalis*) against *Streptococcus Mutans. Brazilian Dental Journal, 23*, 36–38.

52. Saraya, S., Kanta, I., Sarisuta, N., Temsiririrkkul, R., Suvathi, Y., & Samranri K., et al., (2008). Development of Guava Extract Chewable Tablets for Anticariogenic Activity against streptococcus Mutans. *Mahidol University Journal of Pharmaceutical Sciences*, *35*(1-4), 18–23.

53. Schwendicke, F., Dörfer, C. E., Schlattmann, P., Page, L. F., Thomson, W. M., & Paris, S. (2015). Socioeconomic Inequality and Caries: A Systematic Review and Meta-Analysis. *Journal of Ddental Research*, *94*(1), 10–18.

54. Segura, A., Boulter, S., Clark, M., Gereige, R., Krol, D. M., Mouradian, W., et al., (2014). Maintaining and improving the oral health of young children, section on oral, health. *Pediatrics, 134*(6), 1224–1229.

55. Sharma, S., Agarwal, S. S., Prakash, J., Pandey, M., & Singh, A., (2014). Formulation Development and Quality Evaluation of Polyherbal Toothpaste "Oral S". *International Journal of Pharmaceutical Research & Allied Sciences, 3*(2), 30–39.

56. Silk, H., (2014). Diseases of the mouth. *Primary Care, 41*(1), 75–90.

57. Silverstone, L. M., (1973). The structure of carious enamel, including the early lesion. *Oral Science Review*, *3*, 100–60.

58. Silverstone, L. M., Wefel, J. S., Zimmerman, B. F., Clarkson, B. H., & Featherstone, M. J., (1981). Remineralization of natural and artificial lesions in human dental enamel in vitro. *Caries Research*, *15*, 138–157

59. Somu, C. A., Ravindra, S., Soumya, A., & Ahamed, M. G., (2012). Efficacy of a herbal extract gel in the treatment of gingivitis: A clinical study. *Journal of Ayurveda and Integrative Medicine*, *3*(2), 85–90.

60. Southam, J. C., & Soames, J. V., (1993). Dental Caries. In: Oral Pathology. 2nd Ed., Oxford University Press, Oxford,

61. Srivastava, S., Lal, V. K., & Pant, K. K., (2012). Polyherbal formulations based on Indian medicinal plants as antidiabetic phytotherapeutics. *Phytopharmacology, 2*(1), 1–15.

62. Tatikonda, A., Debnath, S., Chauhan, V. S., Chaurasia, V. R., Taranath, M., & Sharma, A. M. (2014). Effects of herbal and non-herbal toothpastes on plaque and gingivitis: A clinical comparative study. *Journal of International Society of Preventive & Community Dentistry, 4* (Suppl 2), S126–S129.

63. Taylor, G. W., (2004). Diabetes, periodontal diseases, dental caries, and tooth loss: a review of the literature. *Compendium of Continuing Education in Dentistry, 25,* 179–188.

64. Thombre, R., Khadpekar, A., & Phatak, A., (2012). Anti-bacterial activity of various medicinal plants against mixed dental flora. *Research Journal of Pharmaceutical, Biological and Chemical Sciences, 3*(3), 179–182.

65. Vanka, A., Tandon, S., Rao, S. R., Udupa, N., & Ramkumar, P., (2001). The effect of indigenous Neem *Azadirachtaindica* [correction of (*Adirachtaindica*)] mouth wash on streptococcus mutans and lactobacilli growth. *Indian Journal of Dental Research, 12*(3), 133–144.

66. Vasconcelos, L. C., Sampaio, F. C., Sampaio, M. C., Pereira, M. S., Higino, J. S., & Peixoto, M. H., (2006). Minimum inhibitory concentration of adherence of *Punicagranatum* Linn (pomegranate) gel against *S. mutans, S. mitis* and *C. albicans. Brazilian Dental Journal, 17*(3), 223–227.

67. Vohra, K., Sharma, M., & Guarve, K., (2012). Evaluation of Herbal toothpowder and its comparison with various marketed toothpaste brands. *International Journal of Green and Herbal Chemistry. 1*(3), 271–276.

68. Watt, R., & Sheiham, A., (1999). Health policy: Inequalities in oral health: a review of the evidence and recommendations for action. *British Dental Journal, 187,* 6–12.

69. Watt, R. G., Listl, S., Peres, M. A., & Heilmann, A., (2015). Social Inequalities in Oral Health: From Evidence to Action. Watt, R. G., Listl, S., Peres, M. A., & Heilmann, A. (Eds.). International Centre for Oral Health Inequalities Research & Policy (ICOIRP): London.

70. WHO, (2016). Oral Health, Fact Sheet No-318, World Health Organization:Geneva, Switzerland, (2012). Accessed on 26 April URL :http://www.who.int/mediacentre/factsheets/fs318/en/

71. Wright, J. T., & Hart, T. C., (2002). The genome projects: Implications for dental practice and education. *Journal of Dental Education, 66,* 659–671.

CHAPTER 3

ENGINEERING INTERVENTIONS FOR EXTRACTION OF ESSENTIAL OILS FROM PLANTS

ASAAD REHMAN SAEED AL-HILPHY

CONTENTS

3.1 INTRODUCTION

Essential oils or etheric oils mean volatile oils and are obtained from plants by steam distillation method [19, 29]. Essential oils are used for medicinal and pharmaceutical purposes, food and food ingredients, herbal tea, cosmetics, perfumery, aromatherapy, pest and disease control, gelling agents, dying in fabrics, plant growth regulators and paper making, etc. Munir and Hensel [28] indicated that essential oils have been used in the medicinal and pharmaceutical purposes, as well as food industries, cosmetics, perfumes, physical therapy, the struggle of insects, diseases treatment, dye fabrics and jellies processing, plant growth regulator and manufacturing paper.

Malle and Schmickl [26] stated that the advantages of distillation methods are extracting pure and refine essential oils by evaporating the volatile essence from the plant. Essential oils can be extracted from all plants or different parts of the plant like bark, leaves, roots, wood, seeds or fruits, flowers, burgeons, branches [29]. Also, Alhakeem and Hassan [4] mentioned that essential oils are extracted from various parts of plants such as roots, stems, leaves, buds, fruits, flowers, seeds and bark. About 65% of the essential oils are produced from the woody plants such as trees and bushes [8]. Herbal products are marketed as fresh, dry products, and essential oils. In general, the plants are used as raw or dried materials for the extraction of essential oils [35]. Al-Hilphy developed and evaluated his own steam equipment to extract the essential oils [7].

This chapter discusses engineering interventions for extraction of essential oils from plants.

3.2 EXTRACTION METHODS FOR ESSENTIAL OILS

3.2.1 HYDRO DISTILLATION

Hydro distillation is used to isolate essential oil from the aromatic plant via boiling water and plants or using steam. Due to the effect of hot water or steam, the essential oils are separated from the oil glands, which are present in the plant tissue. Separated water and oil (vapor mixture) go to the condenser for conversion to liquid and then is transferred to the separator for separating essential oil from water.

3.2.2 PHYSIOCHEMICAL PROCESS DURING HYDRO DISTILLATION MECHANISM

3.2.2.1 Hydro-Diffusion

Hydro-diffusion is a diffusion of hot water (water distillation method) and essential oils through aromatic plant membranes, contrary to steam distillation method in which the dry steam cannot penetrate the dry cell membrane. Therefore, the aromatic plant must be milled when it is distilled by steam distillation method because the essential oils are free after comminution process. The another method for improving steam distillation method is soaking aromatic plant material in the water because plants cell membranes are impermeable to essential oils, and when soaking plants materials in water makes plants cell membranes to be permeable. Also, boiling water causes liquefaction of essential oil in the water inside the glands. In this process, the solution of oil-water permeate plants cell membranes via osmosis and go out of the membrane, the vaporized oil is transferred with steam. On the other hand, the speed of essential oil vaporization is affected by its degree of solubility in water and is not affected by oil components volatility. The time to distillate milled plant material is less than that for the non-milled plant [32].

3.2.2.2   Hydrolysis

Hydrolysis is a chemical reaction between components of essential oils and water. At high temperatures, the esters (essential oils constituents) incline to react with water to produce alcohol and acids but the reaction is not complete in all directions. Increasing the amount of water (in water distillation method) leads to increase the amount of the alcohol and acid and as a result essential oil yield is decreased. Hydrolysis depends on the time of contact between water and oil, and the hydrolysis increases with increasing of distillation time [32].

3.2.2.3   Decomposition by Heat Treatment

At high temperatures, all essential oils constituents are unstable. For improving oil quality, the hydro-distillation temperature must be low. Hydro-diffusion, hydrolysis and decomposition by heat occur at the same time and affect one another. Hydro-distillation is a common traditional extraction method. There are three types of hydro-distillation method for extracting of essential oils from plants.

3.2.3   *WATER DISTILLATION*

Water distillation is used to extract of essential oils from raw or dried plants by diffusion mechanism as shown in the schematic diagram by Sen et al. [38]. The plants are soaked in the container, which has water for preventing overheating and charring of the plants, and then heating water with plants till the steam comes out. The oil comes out and it goes to the condenser where the oil and water are collected in separation flasks. The oil collected in the top layer of hydrosol can be isolated. In this method, the extraction temperature always is below 100°C at the surface of the plants to avoid the evaporation of water and oil together [29, 32]. Heating systems in the extraction of essential oils using water distillation are direct fire, steam jacket, closed steam jacket, closed or open steam coil. The flow diagram of water distillation process has been illustrated by Lawrence et al. [25] and Öztekin et al. [29].

3.2.3.1   Advantages [20]

- It is widely used in the world.
- Water distillation method is inexpensive and easy to construct. It is proper for field operation.
- Boiling water causes motion of plant into distilling flask, which leads to improvement heat transfer.
- There is a direct contact between plant and boiling water.

3.2.3.2   Disadvantages [20]

- Complete extraction is not possible.
- Oil ingredients such as esters are sensitive to hydrolysis while other compounds like acyclic monoterpene hydrocarbons and aldehydes are susceptive to polymerization (water pH is mostly reduced during distillation result in readily hydrolytic reactions).
- Oxygenated ingredients like phenols have a tendency to liquefy in the distilled water, as a result water distillation is not able to removal them completely.
- Water distillation takes a long time to accumulate much oil, as a result the good and bad quality oil are mixed.
- Water distillation is a slow process. On the other hand, it needs much fuel, large area and experience.

- It requires large number of stills.
- Produced oil quality is lower.

3.2.4 MATHEMATICAL MODELING OF ESSENTIAL OIL EXTRACTION

The model of second order mechanism is used to calculate the concentration of oil extracted at any time [31]. Dissolution rate for the oil contained in the solid to solution can be calculated as follows:

$$\frac{dc_s}{dt} = k(C_s - C_t)^2 \tag{1}$$

where, k is the rate constant of the second order extraction (ml/g min.), C_t is the oil concentration in the solution at any time (g/l), C_s is the oil concentration at saturation (g/l), t is the time (min.).

The integrated rate for the second order extraction has been obtained by considering the boundary condition at $t = 0$ to t and $C_t = 0$ to C_t.

$$C_t = \frac{C_s^2 kt}{(1 + C_s kt)^2} \tag{2}$$

By transforming Eq. (2) to linear equation, we get:

$$\frac{t}{C_t} = \frac{1}{C_s^2 k} + \frac{t}{C_s} \tag{3}$$

Extraction rate can be expressed as:

$$\frac{C_t}{t} = \frac{1}{\left(\frac{1}{C_s^2 k} + \frac{t}{C_s}\right)} \tag{4}$$

The initial extraction rate (h), when t is close to 0, can be written as:

$$h = C_s^2 k \tag{5}$$

The concentration of solute at any time can be obtained after rearrangement as:

$$C_t = \frac{t}{\left(\frac{1}{h} + \frac{t}{C_s}\right)} \tag{6}$$

where, h, C_s, k can be determined experimentally from the slope and intercept by plotting $[\frac{t}{C_t}]$ versus t.

3.2.5 OHMIC HEATED–WATER DISTILLATION (OHWD)

Ohmic heating (Joule heating) is a novel thermal treatment. In this technology, internal heat generates in food via passing of alternative electric current within the food, then food converts to electrical resistance [39]. In addition, Ohmic-hydro-distillation is an advanced hydro-distillation technique using ohmic heating process and could be considered as a novel method for the extraction of essential oils. As well as, the results of this study introduced a verdant technology because of less power required per ml of essential oil extraction [14]. Heat generates in the internal of food, in addition, the temperature of heated food is lower than the wall. The control of treatment uniformity requires best modeling inputs. The initial process is varied according to food type. Electric and thermal characteristics are varied during operation.

3.2.5.1 Advantages of Ohmic Heating [17, 44]

- Ohmic heating has a high energy efficiency and volumetric heating.
- Particles temperature is higher than liquid by the same factor of conductivity.
- Decreased fouling.
- Can be done in case solid-liquid food mixture.
- Safe technology; and the product treated by ohmic heating has a high quality.
- Rapid and relatively homogenous heating.
- Alternative voltage is implemented to the electrodes at both ends of food.
- Heating rate depends on the electrical field intensity square and the electric conductivity.
- The electric field intensity had varied with change the distance between electrodes and the applied voltage.

- Formation of fouling on the electrodes when used high voltage for milk pasteurization by ohmic heating because whey proteins denaturation during ohmic heating at high voltage.

3.2.5.2 Disadvantages of Ohmic Heating [18]

- Part of treated food has a high temperature, but the other parts have a low temperature because the food composition is complex.
- A corrosion happens in the electrodes and it needs cleaning continuously.
- Capital investment is higher than the other traditional methods.
- Some processes need pretreatment like blanching.

Al-Hilphy [5, 7] designed a new distiller for essential oils by using ohmic heating (Figure 3.1). It consists of two electrodes and the dimensions of each electrode are 0.075×0.14 m, which are manufactured from stainless steel-316, inner cylinder made of heat plastic, its diameter and

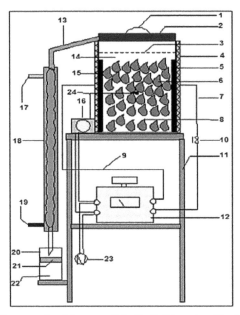

FIGURE 3.1   Components of the essential oils distiller using Ohmic heating by Al-Hilphy [5]. (Reprinted from Al-Hilphy, A. R. S., (2014). Practical study for new design of essential oils extraction apparatus using ohmic heating. *International Journal of Agricultural Science, 4*(12): 351–366. Open access.)

height are 0.14 m and 0.195 m, respectively, provided with the double jacket. Its optimum capacity was 150 g of dried aromatic plants. Glass condensers, a container for condensate water and oil and voltage regulator are also shown in Figure 3.1. The distillation temperature in this apparatus is 100°C. Figure 3.2 illustrates the change in temperature during extraction time at different voltages. The required time to evaporate essential oils with vapor was decreased with the increase in voltage.

Figure 3.3 illustrates that the essential oil yield (%) is directly proportional to extraction time [5]. However, the relationship between the yield of essential oil and extraction time follows a second-degree polynomial. The required time to essential oil distillation was 110, 90, and 142 minutes for ohmic heating hydro (water) distillation (OHHD) at 60, 70, 80 V and HD, respectively. This may be attributed to the higher efficiency of OHHD. The internal heating rate in the case of Ohmic heating is higher. Therefore, its ability to generate heat is high [18, 37]. Also, it can be seen in Figure 3.3 that the percentage of essential oil yield was 1.648, 1.894, 2.177, and 1.369% by using OHHD at 60, 70, 80 V and hydro (water) distillation, respectively.

The system performance coefficient is given by Eq. (7) [21, 22]:

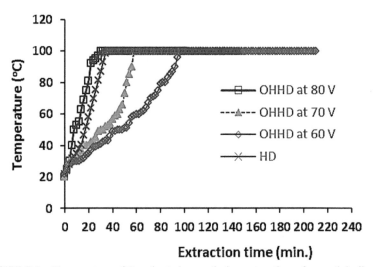

FIGURE 3.2   Temperature of Eucalypts leaves during extraction of essential oil with a distillatory, by Al-Hilphy [5]. (Reprinted from Al-Hilphy, A. R. S., (2014). Practical study for new design of essential oils extraction apparatus using ohmic heating. *International Journal of Agricultural Science, 4*(12): 351–366. Open access.)

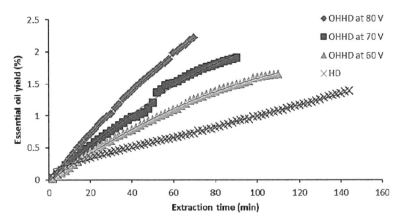

FIGURE 3.3 Essential oil yield extracted from eucalyptus leaves by OHHD and HD methods, by Al-Hilphy [5]. (Reprinted from Al-Hilphy, A. R. S., (2014). Practical study for new design of essential oils extraction apparatus using ohmic heating. *International Journal of Agricultural Science, 4*(12): 351–366. Open access.)

$$spc = \frac{Qt}{Eg} \tag{7}$$

$$Eg = Qt + E_{loss} = \sum[\Delta VIt] \tag{8}$$

$$Q_t = m\,C_p(T_f - T_i) \tag{9}$$

where, m is plant mass (kg), $(T_f - T_i)$ is the difference between extraction temperature and initial temperature (°C), E_g is the given energy (J), Q_t is the heat extracted (J), *SPC* is the System performance coefficient, and C_p is the specific heat (J/kg.K) at constant pressure.

The specific heat can be calculated from the following empirical equation [40]:

$$cp = 4176.2 - 9.0864 \times 10^{-3}\,T + 5473.1 \times 10^{-6}T_2 \tag{10}$$

Heat loss to the ambient is shown below:

$$E_{loss} = h\pi DL(T_w - T_{amb})\Delta t \tag{11}$$

where, h is the heat transfer coefficient (W/m²K), D is the cylinder diameter (m), T_w is the outer cylinder wall temperature (°C), T_{amb}, is the ambient temperature (°C), E_{loss} is the heat loss via natural convection (J), and t is the time.

Average heat transfer coefficient is calculated [15] as follows:

$$h=\left[\frac{\Delta t}{D}\right]^{1/4} \qquad (12)$$

where, ΔT is the temperature difference between the final and the ambient temperature of wall.

It can be seen in Figure 3.4 that the relationship between SPC and voltage gradient is linear (first order) as follows:

$$SPC = -0.007E + 1.0292 \qquad (13)$$

where, E is the voltage gradient.

3.2.6 EFFECTS OF OHMIC HEATING ON SPECIFIC GRAVITY AND REFRACTIVE INDEX OF ESSENTIAL OILS

The specific gravity and reflective index for essential oils was not affected significantly by extraction methods (OHWD at 60, 70, 80 V and WD) as

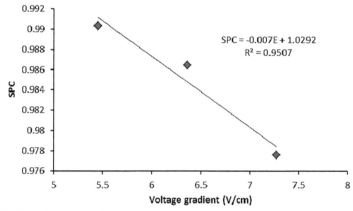

FIGURE 3.4 System performance coefficient versus voltage gradient for Eucalyptus leaves, by Al-Hilphy [5]. (Reprinted from Al-Hilphy, A. R. S., (2014). Practical study for new design of essential oils extraction apparatus using ohmic heating. *International Journal of Agricultural Science, 4*(12): 351–366. Open access.)

shown in Table 3.1 by Al-Hilphy [5]. Alhakeem and Hasan [4] demonstrated that the reflective index and specific gravity of extracted Eucalyptus oil fluctuated between 1.4631–1.4644 and 0.9160–0.9300. Moreover, the reflective index and specific gravity of extracted Eucalyptus oil by water distillation method were 1.4928 and 0.9162, respectively [1]. Therefore, the extracted essential oils from Eucalyptus (*Eucalyptus tereticornis*) leaves by using ohmic heating is an extraction technology that extracts the essential oils with high quality.

3.2.7 EFFECTS OF OHMIC HEATING (OHWD) ON CHEMICAL COMPOSITION OF ESSENTIAL OILS

It Tables 3.2–3.5 indicate the essential oils ingredients of Eucalyptus leaves, which were analyzed by GC-MS [5]. The total components of essential oils were 29, 30, 27 and 28% by using OHWD at 60, 70, 80 V, and WD, respectively. The main component of Eucalyptus essential oil is Eucalyptol. Eucalyptol content in Eucalyptus essential oil was 12.54, 15.66, 39.49, and 15.96% by using OHWD at 60, 70, 80 V, and WD, respectively. Other main ingredients of Eucalyptus essential oil extracted by using OHWD at 60, 70, 80 V and WD were: {α-Pinene (6.07%), Benzene,1-methyl-3-(1-methylethyl)-(22.09%), 1H-Cycloprop[e]azulen-7-ol,decahydro-1,1(7.39%) and (–)-Globulol (5.66%)}, {α-Pinene (6.37%), Benzene, 1-methyl-3-(1-methylethyl)-(21.65%), 3-Allyl-6-methoxyphenol(6.53%), 2-Naphthalenemethanol, decahydro-α,(8.69%)}, {α-Pinene(7.46), Bicyclo[3.1.1] heptan-3-ol, 6,6-dimethyl-2-(5.27%), 3-Allyl-6-methoxyphenol(9.46%), 2-Naphthalenemethanol, decahydro-α,.(5.83%)}, {α-Pinene(11.40%), Benzene, 1-methyl-3-(1-methylethyl)-(5.16%), Bicyclo[3.1.1]heptan-3-ol,

TABLE 3.1 The Reflective Index and Specific Gravity of Eucalyptus Essential Oils Extracted by Ohmic Heating, by Al-Hilphy [5]. (Reprinted from Al-Hilphy, A. R. S., (2014). Practical study for new design of essential oils extraction apparatus using ohmic heating. *International Journal of Agricultural Science, 4*(12): 351–366. Open access.)

Water distillation methods	Specific Gravity	Reflective Index
OHWD 60 V	0.914[a] ± 0.0022	1.4959[a] ± 0.0012
OHWD 70 V	0.913[a] ± 0.0041	1.4945[a] ± 0.0011
OHWD 80 V	0.913[a] ± 0.0033	1.4946[a] ± 0.0014
WD	0.914[a] ± 0.0051	1.4998[a] ± 0.0016

[a] Significant at 5%.

TABLE 3.2 Chemical Compounds of Eucalypts Oil, which was extracted by OHWD at 60 V, by Al-Hilphy [5]. (Reprinted from Al-Hilphy, A. R. S., (2014). Practical study for new design of essential oils extraction apparatus using ohmic heating. *International Journal of Agricultural Science, 4*(12): 351–366. Open access.)

No.	Chemical compound	R.T. (min)	Formula	Area (%)	Mol. Weight
1	α-Pinene	5.255	$C_{10}H_{16}$	6.07	136
2	Benzene, 1-methyl-3-(1-methylethyl)-	7.067	$C_{10}H_{14}$	3.67	134
3	Benzene, 1-methyl-3-(1-methylethyl)-	7.267	$C_{10}H_{14}$	22.09	134
4	Eucalyptol	7.373	$C_{10}H_{18}O$	12.54	154
5	1,4-Cyclohexadiene, 1-methyl-4-(1-methyle	7.716	$C_{10}H_{16}$	1.42	136
6	1,6-Octadien-3-ol, 3,7dimethyl-	8.454	$C_{10}H_{18}O$	0.69	154
7	Butanoic acid, 3-methyl-, 3-methyl-butyl este	8.559	$C_{10}H_{20}O_2$	0.45	172
8	Bicyclo[3.1.1]heptan-3-ol, 6,6-dimethyl-2-m	9.168	$C_{10}H_{16}O$	4.11	152
9	2(10)-Pinen-3-one, (+/–)	9.495	$C_{10}H_{14}O$	1.00	150
10	Borneol	9.660	$C_{10}H_{18}O$	0.68	154
11	3-Cyclohexen-1-ol, 4-methyl-1-(1-methyleth	9.809	$C_{10}H_{18}O$	2.68	154
12	Bicyclo[3.2.0]heptan-3-ol, 2-methylene-6,6-	9.922	$C_{10}H_{16}O$	1.88	152
13	3-Cyclohexene-1-methanol, α,α,4	10.067	$C_{10}H_{18}O$	2.59	154
14	Cyclohexanol, 2-methylene-5-(1-methylethe	10.591	$C_{10}H_{16}O$	1.00	152
15	1-Cyclohexene-1-carboxaldehyde, 4-(1-meth	11.311	$C_{10}H_{16}O$	0.95	152
16	Phenol, 2-methyl-5-(1-methylethyl)-	11.649	$C_{10}H_{14}O$	0.80	150
17	1H-Cycloprop[e]azulene, decahydro-1,1,7-t	13.585	$C_{15}H_{24}$	1.17	204

TABLE 3.2 (Continued)

No.	Chemical compound	R.T. (min)	Formula	Area (%)	Mol. Weight
18	1H-Cycloprop[e]azu-lene, decahydro-1,1,7-tr	13.875	$C_{15}H_{24}$	1.58	204
19	Oxalic acid, mono-amide, N-(2-phenyl-ethyl)	14.257	$C_{16}H_{23}NO_3$	0.91	277
20	1H-Cycloprop[e] azulen-7-ol, deca-hydro-1,1	15.458	$C_{15}H_{24}O$	7.39	220
21	(−)-Globulol	15.560	$C_{15}H_{26}O$	5.66	222
22	Cubenol	15.644	$C_{15}H_{26}O$	1.28	222
23	2-Naphthalenemetha-nol, 2,3,4,4a,5,6,7,8-oct	15.974	$C_{15}H_{26}O$	0.45	222
24	2-Naphthalenemetha-nol, 1,2,3,4,4a,5,6,7-oc	16.067	$C_{15}H_{26}O$	0.62	222
25	2-Naphthalenemetha-nol, decahydro-α	16.375	$C_{15}H_{26}O$	3.85	222
26	Spiro[5.5]undec-2-ene, 3,7,7-trimethyl-11-m	17.296	$C_{15}H_{24}$	3.96	204
27	1,2-Benzenedicarbox-ylic acid, bis(2-methyl	18.640	$C_{16}H_{22}O_4$	3.34	278
28	1,2-Benzenedicar-boxylic acid, butyl 2-methy	19.619	$C_{24}H_{38}O_4$	1.37	390
29	1,2-Benzenedicarbox-ylic acid	24.889	$C_{16}H_{22}O_4$	5.80	278

TABLE 3.3 Chemical Compounds of Eucalypts Oil extracted by OHWD at 70 V, by Al-Hilphy [5]. (Reprinted from Al-Hilphy, A. R. S., (2014). Practical study for new design of essential oils extraction apparatus using ohmic heating. *International Journal of Agricultural Science, 4*(12): 351–366. Open access.)

No.	Chemical compound	R.T. (min)	Formula	Area (%)	Mol. Weight
1	α-Pinene	5.247	$C_{10}H_{16}$	6.37	136
2	Bicyclo[3.1.1]heptane, 6,6-dimethyl-2-meth	6.096	$C_{10}H_{16}$	0.44	136
3	Benzene, 1-methyl-3-(1-methy-lethyl)-	7.056	$C_{10}H_{14}$	2.68	134

TABLE 3.3 (Continued)

No.	Chemical compound	R.T. (min)	Formula	Area (%)	Mol. Weight
4	Benzene, 1-methyl-3-(1-methylethyl)-	7.258		21.65	
5	Eucalyptol	7.363	$C_{10}H_{18}O$	15.66	154
6	1,4-Cyclohexadiene, 1-methyl-4-(1-methyle	7.715	$C_{10}H_{16}$	1.69	136
7	(+)-4-Carene	8.181	$C_{10}H_{16}$	0.88	136
8	Bicyclo[3.1.1]heptan-3-ol, 6,6-dimethyl-2-	9.160	$C_{10}H_{16}O$	4.25	152
9	2(10)-Pinen-3-one, (+/–)	9.486	$C_{10}H_{14}O$	0.95	150
10	Bicyclo[2.2.1]heptan-2-ol, 1,7,7-trimethyl-,	9.642	$C_{10}H_{18}O$	0.41	154
11	3-Cyclohexen-1-ol, 4-methyl-1-(1-methylet	9.786	$C_{10}H_{18}O$	1.92	154
12	Cyclohexanol, 2-methylene-5-(1-methylethe	9.930	$C_{10}H_{16}O$	1.32	152
13	3-Cyclohexene-1-methanol, α,α,4	10.044	$C_{10}H_{18}O$	1.62	154
14	Cyclohexanol, 2-methylene-5-(1-methylethe	10.578	$C_{10}H_{16}O$	0.91	152
15	3-Allyl-6-methoxyphenol	12.458	$C_{10}H_{12}O_2$	6.53	164
16	Caryophyllene	13.323	$C_{15}H_{24}$	0.37	204
17	1H-Cycloprop[e]azulene, decahydro-1,1,7-t	13.586	$C_{15}H_{24}$	1.46	204
18	1H-Cycloprop[e]azulene, decahydro-1,1,7-t	13.877	$C_{15}H_{24}O$	2.02	220
19	β-Guaiene	14.248	$C_{15}H_{24}$	0.69	204
20	2-Isopropenyl-4a,8-dimethyl-1,2,3,4,4a,5,6,	14.342	$C_{15}H_{24}$	0.40	204
21	Phenol, 2-methoxy-4-(2-propenyl)-, acetate	14.591	$C_{12}H_{14}O_3$	0.98	206
22	Cyclohexanemethanol, 4-ethenyl-α,.al	15.033	$C_{15}H_{26}O$	2.36	222
23	1H-Cycloprop[e]azulen-7-ol, decahydro-1,1	15.430	$C_{15}H_{24}O$	4.83	220
24	(–)-Globulol	15.530	$C_{15}H_{26}O$	4.50	222
25	(–)-Globulol	15.620	$C_{15}H_{26}O$	0.92	222
26	Ledol	15.737	$C_{15}H_{26}O$	0.49	222

TABLE 3.3 (Continued)

No.	Chemical compound	R.T. (min)	Formula	Area (%)	Mol. Weight
27	2-Naphthalenemethanol, 1,2,3,4,4a,5,6,7-oct	16.071	$C_{15}H_{26}O$	1.65	222
28	2-Naphthalenemethanol, decahydro-α	16.404	$C_{15}H_{26}O$	8.69	222
29	γ-Neoclovene	17.277	$C_{15}H_{24}$	2.76	204
30	1,2-Benzenedicarboxylic acid, bis(2-methyl	18.597	$C_{16}H_{22}O_4$	0.61	278

TABLE 3.4 Chemical Compounds of Eucalypts Oil Extracted by OHWD at 80 V, by Al-Hilphy [5]. (Reprinted from Al-Hilphy, A. R. S., (2014). Practical study for new design of essential oils extraction apparatus using ohmic heating. *International Journal of Agricultural Science, 4*(12): 351–366. Open access.)

No.	Chemical compound	R.T. (min)	Formula	Area (%)	Mol. Weight
1	α-Pinene	5.255	$C_{10}H_{16}$	7.46	136
2	Bicyclo[3.1.1]heptane, 6,6-dimethyl-2-meth	6.098	$C_{10}H_{16}$	0.54	136
3	α-Phellandrene	6.689	$C_{10}H_{16}$	0.70	136
4	Benzene, 1-methyl-3-(1-methy-lethyl)-	7.054	$C_{10}H_{14}$	2.04	134
5	Eucalyptol	7.379	$C_{10}H_{18}O$	39.49	154
6	1,4-Cyclohexadiene, 1-methyl-4-(1-methyle	7.714	$C_{10}H_{16}$	1.42	136
7	(+)-4-Carene	8.182	$C_{10}H_{16}$	0.80	136
8	Bicyclo[3.1.1]heptan-3-ol, 6,6-dimethyl-2-	9.173	$C_{10}H_{16}O$	5.27	152
9	2(10)-Pinen-3-one, (+/−)	9.497	$C_{10}H_{14}O$	1.39	150
10	Isoborneol	9.646	$C_{10}H_{18}O$	0.53	154
11	3-Cyclohexen-1-ol, 4-methyl-1-(1-methylet	9.788	$C_{10}H_{18}O$	1.60	154
12	Cyclohexanol, 2-methylene-5-(1-methylethe	9.931	$C_{10}H_{16}O$	1.42	152
13	3-Cyclohexene-1-methanol, α,α,4	10.042	$C_{10}H_{18}O$	1.35	154
14	Cyclohexanol, 2-methylene-5-(1-methylethe	10.581	$C_{10}H_{16}O$	1.04	152
15	3-Allyl-6-methoxyphenol	12.481	$C_{10}H_{12}O_2$	9.46	164

TABLE 3.4   (Continued)

No.	Chemical compound	R.T. (min)	Formula	Area (%)	Mol. Weight
16	1H-Cycloprop[e]azulene, decahydro-1,1,7-t	13.589	$C_{15}H_{24}$	1.68	204
17	1H-Benzocycloheptene, 2,4a,5,6,7,8-hexahy	14.249	$C_{15}H_{24}$	0.79	204
18	1,5-Cyclodecadiene, 1,5-dimethyl-8-(1-met	14.342	$C_{15}H_{24}$	0.41	204
19	Phenol, 2-methoxy-4-(2-prope-nyl)-, acetate	14.607	$C_{12}H_{14}O_3$	2.00	206
20	Cyclohexanemethanol, 4-ethenyl-α,.al	15.022	$C_{15}H_{26}O$	1.19	222
21	Epiglobulol	15.192	$C_{15}H_{26}O$	0.53	222
22	1H-Cycloprop[e]azulen-7-ol, decahydro-1,1	15.425	$C_{15}H_{24}O$	4.50	220
23	(–)-Globulol	15.530	$C_{15}H_{26}O$	4.14	222
24	2-Naphthalenemethanol, 1,2,3,4,4a,5,6,7-oct	16.066	$C_{15}H_{26}O$	1.19	222
25	2-Naphthalenemethanol, decahydro-α	16.381	$C_{15}H_{26}O$	5.83	222
26	γ-Neoclovene	17.272	$C_{15}H_{24}$	2.52	204
27	1,2-Benzenedicarboxylic acid, bis(2-methyl	18.598	$C_{16}H_{22}O_4$	0.71	278

TABLE 3.5 Chemical Compounds of Eucalypts Oil Extracted by Water Distillation (WD), by Al-Hilphy [5]. (Reprinted from Al-Hilphy, A. R. S., (2014). Practical study for new design of essential oils extraction apparatus using ohmic heating. *International Journal of Agricultural Science, 4*(12): 351–366. Open access.)

No.	Chemical compound	R.T. (min)	Formula	Area (%)	Mol. Weight
1	α-Pinene	5.279	$C_{10}H_{16}$	11.40	136
2	Benzene, 1-methyl-3-(1-methyle-thyl)-	7.063	$C_{10}H_{14}$	5.16	134
3	Eucalyptol	7.364	$C_{10}H_{18}O$	15.96	154
4	1,4-Cyclohexadiene, 1-methyl-4-(1-methyle	7.708	$C_{10}H_{16}$	1.46	136
5	Bicyclo[3.1.1]heptan-3-ol, 6,6-dimethyl-2-	9.213	$C_{10}H_{16}O$	10.25	152
6	2(10)-Pinen-3-one, (.+/-.)-	9.522	$C_{10}H_{14}O$	3.42	150
7	Bicyclo[2.2.1]heptan-2-ol, 1,7,7-trimethyl-,	9.657	$C_{10}H_{18}O$	1.08	154

TABLE 3.5 (Continued)

No.	Chemical compound	R.T. (min)	Formula	Area (%)	Mol. Weight
8	3-Cyclohexen-1-ol, 4-methyl-1-(1-methylet	9.787	$C_{10}H_{18}O$	1.21	154
9	Cyclohexanol, 2-methylene-5-(1-methylethe	9.934	$C_{10}H_{16}O$	1.38	152
10	3-Cyclohexene-1-methanol, $\alpha,\alpha,4$	10.042	$C_{10}H_{18}O$	1.35	154
11	Cyclohexanol, 2-methylene-5-(1-methylethe	10.584	$C_{10}H_{16}O$	1.09	152
12	2-Cyclohexen-1-one, 2-methyl-5-(1-methyle	10.792	$C_{10}H_{14}O$	0.83	150
13	3-Allyl-6-methoxyphenol	12.504	$C_{10}H_{12}O_2$	12.47	164
14	1H-Cycloprop[e]azulene, decahydro-1,1,7-t	13.607	$C_{15}H_{24}$	4.08	204
15	3-Phenylpropanoic acid, dodec-9-ynyl ester	14.249	$C_{21}H_{30}O_2$	0.70	314
16	Phenol, 2-methoxy-4-(2-propenyl)-, acetate	14.624	$C_{12}H_{14}O_3$	2.97	206
17	Epiglobulol	15.198	$C_{15}H_{26}O$	1.16	222
18	1H-Cycloprop[e]azulen-7-ol, decahydro-1,1	15.418	$C_{15}H_{24}O$	3.15	220
19	(−)-Globulol	15.548	$C_{15}H_{26}O$	6.85	222
20	Cubenol	15.632	$C_{15}H_{26}O$	1.54	222
21	Apiol	15.913	$C_{12}H_{14}O_4$	1.89	222
22	2-Naphthalenemethanol, decahydro-α,	16.353	$C_{15}H_{26}O$	1.90	222
23	Apiol	16.559	$C_{12}H_{14}O_4$	1.08	222
24	4,6,6-Trimethyl-2-(3-methylbuta-1,3-dienyl	17.256	$C_{15}H_{22}O$	1.27	218
25	1,2-Benzenedicarboxylic acid, bis(2-methyl	18.597	$C_{16}H_{22}O_4$	0.54	278
26	12-Oleanen-3-yl acetate, (3α)-	25.709	$C_{32}H_{52}O_2$	1.05	468
27	4,4,6a,6b,8a,11,11,14b-Octa-methyl-1,4,4a,5	26.405	$C_{30}H_{48}O$	3.07	424
28	9, 19-Cyclolanost-24-en-3-ol, (3β)-	27.268	$C_{30}H_{50}O$	1.67	426

6,6-dimethyl-2-(10.25%), 3-Allyl-6-methoxyphenol (12.47%), (−)-Globu-lol(6.85%)}, respectively. The results illustrated that the Eucalyptol content was significantly affected by using ohmic heating. In the case of using

OHWD at 60, 70, 80 V and WD, the oil consisted of 72.41, 56.66, 62.96 and 85.71% of oxygenated monoterpenes ingredients, respectively, and 27.58, 43.33, 37.03 and 14.28% of monoterpenes hydrocarbons, respectively.

3.2.8   MICROWAVE-ASSISTED WATER DISTILLATION (MAWD)

Microwave is an electromagnetic field lies between frequencies of 300 MHz to 300 GHz [24]. *International Telecommunication Union* (ITU) determined 2450 MHZ as the vibration range for heating by microwave. Its characteristics are same as the visible light. The electromagnetic waves can be transferred through the materials without absorption.

Microwave assisted water distillation (MAWD) occurs by electromagnetic waves, which cause structural changes in the cells. Heat and mass transfer are in the same direction, it is from inside to outside, contrary to traditional extraction where mass transfer occurs from outside to inside (from medium heating to the inner of sample), but heat transfer moves from inside to outside. There are two mechanisms of microwave assisted water distillation:

- The ionic polarization, which means that when electromagnetic field is applied on the food solutions which have a lot of ions, the ions move with high speed resulting impact among ions and friction of which lead to conversion of dynamic energy of ions to heat energy, and the movement increases with the increase of ions concentration; and as a result, the temperature increases highly.
- The dipole rotation, which means rearrangement of dipoles with the applied electromagnetic field. Water contains polarity molecules that have randomized rotation. Microwave energy arrives directly to aromatic plants (materials) through molecular interaction with the electromagnetic field via conversion of electromagnetic energy into heat energy [12, 34, 41].

MAWD has been used to extract essential oil from Rosemary (*Rosmarinus officinalis*) [24], *Lippia alba* [40], and Rosmarinus Dwarfed *Cinnamomum camphora* var. *Linaolifera Fujito* [12]. Wei et al. [43] extracted the essential oils of Dwarfed *Cinnamomum comphora* var *Linaolifera Fujita* using microwave-assisted water distillation and compared it with the traditional water distillation. They stated that the time required to extract essential oil using microwave-assisted water distillation and conventional water distillation reached 37.5 and 120 minutes, respectively and oil yield reached 1.73 and 1.71%, respectively. Fadel et al. [13] indicated that the total essential oil yield obtained by using water distillation and microwave-assisted water distillation reached 1.21 and 1.47%, respectively.

Resan [33] manufactured an apparatus for extracting essential oils by using microwave assisted water distillation. It consisted of microwave apparatus with a power and frequency of 1000 W and 2.45 GHz, respectively, glass condenser, separator and flask.

Figure 3.5 shows the temperature variation of essential oils that were extracted from caraway using MAWD and conventional water distillation (TWD). The results illustrated that the extraction temperature of essential oils from caraway was increased with increase in extraction time. The extraction temperature by using microwave-assisted water distillation was higher than that for conventional method. The MAWD gave a highest extraction temperature compared with TWD and reached to 98.65 and 96.21°C, respectively. These variations in the extraction temperatures between both extraction methods are due to the direction of heating. In the case of microwave, the heat transfer will be from inner towards outer and therefore, the heat loss is contrary to TWD.

It can be seen in Figure 3.6 that the oil yield distilled from aniseed and fennel by using MAWD was highest compared with TWD because of increasing heat efficiency [6]. Therefore, randomized ions movement velocity increases which leads to increase in the impact energy among ions, thereby the cells walls, which containing oil, are destroyed and release maximum amount of essential oils in water. As shown in Figure 3.6, oil yield extracted from aniseed by using MAWD and TWD reached 3.7 and 2.8%, respectively. Chemat et al. [10] demonstrated that heat generated by microwave causes variation in the pressure between inner and exterior of plant cells, as well as most compounds are easily released with increasing coefficient of mass transfer and outside cells are destroyed completely. Kapas et al. [23] and Azar [8] indicated that the essential oils extracted by MAWD are higher by 7.5% compared with TWD. The results illustrates that the variation between MAWD and TWD in oil yield of caraway and cumin is not significant.

3.2.8.1   Effects of MAWD on the Physical Properties of Essential Oils

Refractive index of essential oils extracted from caraway, cumin and aniseed by using MAWD is less than TWD, but for Fennel oil it was higher due to increase in number of unsaturated bonds. According to Resan [33], the refractive index of essential oils extracted from Cumin by using MAWD and TWD reached to 1.5056 and 1.5059, respectively, and for Fennel reached to 1.5561 and 1.5449, respectively. Sainin et al. [36] stated that the refractive index of cumin essential oil is 1.4675. Abo-zaid [2] demonstrated that the refractive index of cumin essential oil is 1.4720. Whereas, Abo-zaid [2] indicated that the refractive index of Caraway essential oil range was between 1.4860 - 1.4878. Al-Mayah [3] revealed presence of slow evaporating ingredients in the essential oils as a lot of oxygenated compounds cause an increase of refractive index of essential oils. In addition to the variance in the refractive index, it may be attributed to the presence of lot of unsaturated bonds [3]. Pronpunyapat et al. [30] stated that the increase of refractive index and relative density makes essential oils color as dark.

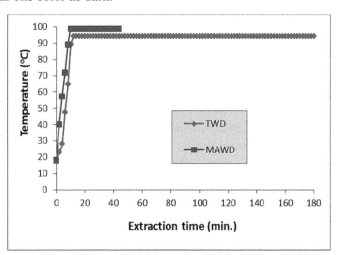

FIGURE 3.5   Temperature of essential oil extraction of caraway by MAWD and TWD, by Al-Hilphy [6]. (Reprinted from Al-Hilphy, A. R. S., Al-fekaiki, D. F., & Hussein, R. A., (2015). Extraction of essential oils from some types of umbelifera family using microwave-assisted water distillation. *Journal of Biology, Agriculture and Healthcare,* 5(22), 16–28. Creative Commons Attribution 3.0 Unported (CC BY 3.0) License. https://creativecommons.org/licenses/by/3.0/)

The viscosity of all essential oil extracted by using MAWD is insignificantly lower than that for TWD except viscosity of Cumin oil has the same value in both methods. There is no significant effect between MAWD and TWD in the specific density of all essential oils. Abo-zaid [2] demonstrated that the specific density of cumin essential oil is 0.887. The specific density of cumin essential oil extracted by TWD and MAWD reached 0.724 and 0.744, respectively. Sainin et al. [36] stated that the specific density of cumin essential oil is 0.7455. Damayanti and Setyawn [11] showed that the specific density of fennel essential oil ranged from 0.978 to 0.988.

3.2.8.2 Effects of MAWD on Chemical Composition of Essential Oils

Resan [32] carried out a study on the identification of active compounds by GC-MS of essential oil extracted from caraway using Clevenger and microwave-assisted water distillation. The researcher demonstrated appearance of Carvon compound by high ratio and it reached 47.12 and 40.5% for both extraction methods, respectively, while compound ratio of Anethol and D-Limonene of caraway oil extracted by Clevenger were 16.18 and 11.67%, respectively, and by microwave-assisted water distillation were 22.19 and 10.96%, respectively.

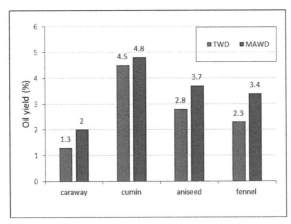

FIGURE 3.6 Oil Yield of Essential Oils from caraway, cumin, anise and fennel seeds extracted by using MAWD and TWD [6]. (Reprinted from Al-Hilphy, A. R. S., Al-fekaiki, D. F., & Hussein, R. A., (2015). Extraction of essential oils from some types of umbelifera family using microwave-assisted water distillation. *Journal of Biology, Agriculture and Healthcare, 5*(22), 16–28. Creative Commons Attribution 3.0 Unported (CC BY 3.0) License. https://creativecommons.org/licenses/by/3.0/)

The chemical compounds (like A-isopropylbenzyl alcohol, Cumic alde-hyde and betapinene) were extracted from essential oil of cumin by using Clevenger; and the values reached 35.19, 22.60 and 13.26%, respectively, while values were 33.77, 21.24 and 13.26%, respectively, by using micro-wave assisted water distillation [32]. On the other hand, the A-Isopropyl-benzyl alcohol compound had higher concentration compared with other prevailing compounds in the extracted oil by these two methods. Results by Resan [32] showed appearance of chemical compounds in aniseed oil extracted by Clevenger and the higher ratio was registered by Anethol com-pound that reached 75.33% compared with other prevailing compounds. Also appearance of the same compound with high ratio reached 77.58% in the extracted oil by microwave-assisted water distillation. The prevailing compounds in the essential oil of fennel extracted were Anethol, L-Fen-chon and Estrago and their ratio was 72.78, 7.41 and 5.52% by Clevenger method, compared to 67.01, 8.28 and 4.93%, respectively by using micro-wave-assisted water distillation. Among these results, the Anethol com-pound was the prevailing compound among the two extraction methods.

3.2.9   STEAM DISTILLATION

Steam distillation (SD) is a widespread method for isolating essential oils commercially. About 80 to 90% of the essential oils are produced by steam distillation method [7]. This method is used for extracting essential oils from fresh plant materials that have a high boiling point such as roots and seeds. Also, the essential oil in the peppermint, spearmint, oil roses and chamomile are extracted by using steam distillation method. In this method, the plant material is placed on the perforated grid, then steam is released from steam boiler to the extraction vat and passing through the plant mate-rial, and as a result the essential oil is separated from plant material by the diffusion process, and comes out with steam to the condenser, and then to the separation unit. The components of SD are shown in (Figure 3.7).

3.2.9.1   Advantages

- It has high energy efficient.
- Cheapest way for extracting essential oils when on a small-scale.

- Essential oils produced by steam distillation have a high quality.
- The control on the distillation rate is better.
- Working pressure can be changed according to the working conditions.
- No decomposition of in oil compounds due to steam.

3.2.9.2 Disadvantages

- Set up cost of a large scale of essential oil extractor by this method is high.

3.2.9.3 Calculation of Extracted Oil Ratio

Ratio of extracted oil explains that extraction of essential oil occurs via washing and diffusion mechanisms as follows [27]:

$$\frac{q}{q_\infty} = 1 - f e - k_1 t - (1-f) e - k_2 t \tag{14}$$

where: q is the quantity of clove oil extracted at time (t), q_∞ is the quantity of clove oil found in clove flowers at 100% extraction, f is the washing factor, k_1 is the washing constant, k_2 is the diffusion constant.

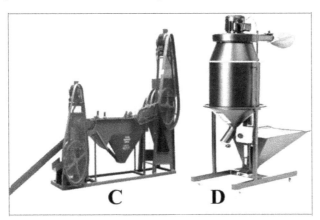

C   D

FIGURE 3.7 Components of steam distillation method for extraction of the essential oils, by Al-Hilphy [7]. (Reprinted from Al-Hilphy, A. R. S., (2015). Development of steam essential oils extractor. *IOSR Journal of Agriculture and Veterinary Science, 8*(12), 2319–2372. © International Organization of Scientific Research (IOSR). Used with permission.)

If washing is very fast and occurs instantaneously, then $k \to \infty$ and we get:

$$\frac{q}{q_\infty} = 1-(1-f)e-k_2t \tag{15}$$

If no washing of the essential oil occurs ($f = 0$), then we have:

$$\frac{q}{q_\infty} = 1-e-k_2t \tag{16}$$

The constants f, k_1, k_2 in Eqs. (14)–(16) are very important to predict [q/q_∞]. The constants f, k_1, k_2 can be calculated by using the worksheet in excel.

3.2.10 DEVELOPMENT OF ESSENTIAL OILS EXTRACTED BY STEAM DISTILLATION

Development of steam extractor for essential oils was carried out by Al-Hilphy [7] and the equipment consisted of following components (Figure 3.8):

- Small boiler.
- *Extraction unit* consisted of a stainless steel cylinder of 120 cm length and 5 cm outside diameter and 2 mm thickness. There is a side slot in the bottom of the cylinder to permit entrance the steam, as well as it has another orifice in the bottom provided with a valve to drain the water from the cylinder. Also, the upper part of the upper cylinder contains an orifice that permits to exit both the steam and oil together. There is a shaft of 125 cm length with slide perforated stainless steel disk that is put inside the cylinder Also, a thermocouple is used to measure of plant temperature.
- Glass condenser is used as a heat exchanger. It contains an internal corrugated glass tube.
- Separator flask of 1 liter is used to isolate the oil layer from water.

3.2.10.1 Distillation Temperature

Figure 3.9 illustrates that clove temperature using developed steam distiller is less than the traditional steam distiller. This is due to the fact

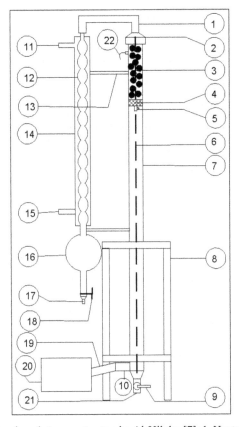

FIGURE 3.8 Developed steam extractor, by Al-Hilphy [7]: 1. Heat plastic pipe, 2. Cover, 3. Clove flowers, 4. Net metal, 5. Slide cylinder, 6. Bar, 7. Cylinder 8. Wood matrix, 9. Valve, 10. Steam inlet, 11. Water outlet, 12. Corrugated pipe, 13. Bar, 14. Outer cylinder, 15. Water inlet, 16. Separation flask, 17. Water out let, 18. Valve, 19. Heat pipe, 20. Steam generator, 21. Water out let, and 22. Thermocouple. (Reprinted from Al-Hilphy, A. R. S., (2015). Development of steam essential oils extractor. *IOSR Journal of Agriculture and Veterinary Science*, 8(12), 2319–2372. © 2015 IOSR.)

that steam loses a part of its energy during the passage in the cylinder by using developed and traditional steam extractor. Mean temperatures were 84.77 and 97.76°C by using developed and traditional steam distiller, respectively. The time required to reach these temperatures was 30 and 15 minutes, respectively. The required total time for complete distillation was 180 and 150 minutes, respectively.

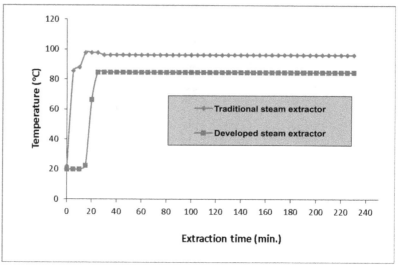

FIGURE 3.9 Relationship for clove oil temperature versus extraction time during the distillation process by using developed and traditional distillers, by Al-Hilphy [7]. (Reprinted from Al-Hilphy, A. R. S., (2015). Development of steam essential oils extractor. *IOSR Journal of Agriculture and Veterinary Science, 8*(12), 2319–2372. © 2015 IOSR.)

3.2.10.2   Effects of the Developed and Traditional Distiller on Physical Properties, Oil Yield and Energy Consumption

Table 3.6 Illustrates that clove oil yield extracted with developed extractor was higher than the clove oil yield extracted by using the traditional extractor. Clove oil yield extracted by the developed and traditional extractor was 13.5 and 11%, respectively. This is due to fact that the increase in oil temperature by using the traditional distiller led to the decomposition of oil compounds that are volatile. The differences in the distilled oil density by developed and traditional distiller were significant.

Table 3.6 shows that the oil viscosity distilled by developed distiller was significantly higher than that with traditional distiller. This is because of increase of oil density distilled by developed distiller. Oil viscosity distilled by using developed and traditional distillers was 0.011029 and 0.00947 Pa-sec, respectively. The differences between the refractive index of oil distilled using developed and traditional distillers were significant. In spite of the energy consumption per ml using developed distiller was lower than traditional distiller, but the differences were not significant.

TABLE 3.6　Oil yield (%), Energy Consumption and Physical Properties of Clove Oil Distilled by Using Developed and Traditional Distillers, by Al-Hilphy [7]. (Reprinted from Al-Hilphy, A. R. S., (2015). Development of steam essential oils extractor. *IOSR Journal of Agriculture and Veterinary Science, 8*(12), 2319–2372. © 2015 IOSR.)

Extractor type	Energy consumption per ml, kW.h/ml	Reflective index	Oil viscosity Pa-sec	Oil density, g/cm^3	Oil yield (%)
Developed	0.1875a	1.5324[a]	0.01029[a]	1.0896[a]	13.5[a]
Traditional	0.1904[a]	1.4509[b]	0.00947[b]	1.0012[b]	11[b]

[a] Significant at 5%;　[b] Significant at 1%.

3.2.10.3　Effect of Developed Distiller on the Chemical Composition of Clove Essential Oil

Tables 3.7 and 3.8 show the chemical compounds in the extracted clove oil by the developed and traditional distiller. The significant chemical compounds in the clove oil are Eugenol (72.69%), 3-Allyl-6-methoxy-phenyl acetate (18.83%), Caryophyllene (3.10%) by using developed distiller and reached to 78.1, 14.85, and 1.89%, respectively by using traditional distiller. On the other hand, the highest percentage of compounds is oxygenated compounds that were reached 95.62 and 97.52% by the developed and traditional distillers, respectively. The percentage of terpene, nitrogen compounds was 4.06, 0.07 and 0.25% by developed distiller, and 2.37 and 0.25% by traditional distiller, respectively. In fact, oil quality had a significant effect due to the distillation methods as illustrated by Al-Hilphy [7], which explains that there is a decomposition in the clove oil when the traditional distillation is used at temperature of 97.76°C, but this problem did not occur when the developed distiller is used because the operation temperature was 84.77°C.

3.3　PROSPECTS

The prospective future of essential oils distillation depends on the improvement of water distillation and steam distillation methods by using new technologies such as ultrasonic treatments, super-critical fluids especially supercritical CO_2, as well as designing systems for distilling essential oils using infrared radiation technology.

TABLE 3.7   Identification of Chemical Compounds Using GC-MS of Clove Oil That was Distillated by Developed Distiller, by Al-Hilphy [7].   (Adapted from Al-Hilphy, A. R. S., (2015). Development of steam essential oils extractor. *IOSR Journal of Agriculture and Veterinary Science, 8*(12), 2319–2372. © 2015 IOSR.)

Mol. Weight	Formula	Area %	R.T. (min)	Name	Peak
136	$C_{10}H_{16}$	0.12	5.138	1R-α-Pinene	1
136	$C_{10}H_{16}$	0.09	5.999	Bicyclo[3.1.1]heptane, 6,6-dimethyl-2-meth	2
134	$C_{10}H_{14}$	0.09	6.915	Benzene, 1-methyl-3-(1-methylethyl)-	3
154	$C_{10}H_{18}O$	1.64	7.037	Eucalyptol	4
172	$C_{10}H_{20}O_2$	0.06	7.243	Acetic acid, sec-octyl ester	5
152	$C_{10}H_{16}O$	0.07	8.985	Bicyclo[2.2.1]heptan-2-one, 1,7,7-trimethyl-	6
233	$C_{13}H_{15}NO_3$	0.07	9.388	Cyclohexanone, 4-(benzoyloxy)-, oxime	7
152	$C_8H_8O_3$	0.27	9.713	Methyl salicylate	8
176	$C_{11}H_{12}O_2$	0.42	10.652	Phenol, 4-(2-prope-nyl)-, acetate	9
164	$C_{10}H_{12}O_2$	72.69	12.114	Eugenol	10
842	$C_{36}H_{74}P_4Pd_2$	0.25	12.293	Palladium(0), bis(.eta.-2-butadiene) 1,1,4,5,8	11
204	$C_{15}H_{24}$	0.17	12.342	Copaene	12
178	$C_{11}H_{14}O_2$	0.07	12.675	Benzene, 1,2-dime-thoxy-4-(2-propenyl)-	13
164	$C_{10}H_{12}O_2$	0.14	12.723	Phenol, 2-methoxy-4-(1-propenyl)-, (E)-	14
204	$C_{15}H_{24}$	3.10	12.906	Caryophyllene	15
164	$C_{10}H_{12}O_2$	0.11	13.283	Phenol, 2-methoxy-4-(1-propenyl)-, (E)-	16
204	$C_{15}H_{24}$	0.49	13.359	1,4,7,-Cycloundecatri-ene, 1,5,9,9-tetramethy	17
206	$C_{12}H_{14}O_3$	18.43	14.149	3-Allyl-6-methoxyphe-nyl acetate	18
220	$C_{15}H_{24}O$	0.25	14.853	1H-Cycloprop[e] azulen-7-ol, deca-hydro-1,1	19
220	$C_{15}H_{24}O$	0.70	14.917	Caryophyllene oxide	20

TABLE 3.7 (Continued)

Mol. Weight	Formula	Area %	R.T. (min)	Name	Peak
220	$C_{15}H_{24}O$	0.11	15.241	12-Oxabicyclo[9.1.0] dodeca-3,7-diene, 1,5,	21
222	$C_{15}H_{26}O$	0.08	15.508	Cubenol	22
220	$C_{15}H_{24}O$	0.17	15.558	Tetracy-clo[6.3.2.0(2,5).0(1,8)] tridecan-9-ol,	23
222	$C_{15}H_{26}O$	0.09	15.765	Cubenol	24
216	$C_{15}H_{20}O$	0.08	15.862	Ar-tumerone	25
346	$C_{22}H_{34}O_3$	0.14	15.940	Kauran-18-al, 17-(acetyloxy)-, (4β)-	26
212	$C_{14}H_{12}O_2$	0.09	17.017	Benzyl Benzoate	27

TABLE 3.8 Identification of Chemical Compounds Using GC-MS of Clove Oil That Was Distillated by Traditional Distiller, by Al-Hilphy [7]. (Adapted from Al-Hilphy, A. R. S., (2015). Development of steam essential oils extractor. *IOSR Journal of Agriculture and Veterinary Science, 8*(12), 2319–2372. © 2015 IOSR.)

M.W.	Formula	Area %	R.T. min	Name	Peak
154	$C_{10}H_{18}O$	0.15	7.034	Eucalyptol	1
148	$C_{10}H_{12}O$	0.30	10.472	Benzaldehyde, 4-(1-methylethyl)-	3
176	$C_{11}H_{12}O_2$	0.33	10.648	Phenol, 4-(2-prope-nyl)-, acetate	4
132	C_9H_8O	0.30	10.941	2-Propenal, 3-phenyl-	5
148	$C_{10}H_{12}O$	0.27	11.110	Benzene, 1-methoxy-4-(1-propenyl)-	6
164	$C_{10}H_{12}O_2$	78.71	12.114	Eugenol	7
790	$C_{46}H_{54}N_4O_8$	0.11	12.158	Dimethyl 2,7,12,18-tet-ramethyl-3,8-di(2,2-d	8
164	$C_{10}H_{12}O_2$	0.55	12.710	Phenol, 2-methoxy-4-(1-propenyl)-, (E)-	10
204	$C_{15}H_{24}$	1.89	12.903	Caryophyllene	11
164	$C_{10}H_{12}O_2$	0.59	13.273	Phenol, 2-methoxy-4-(1-propenyl)-, (E)-	12
204	$C_{15}H_{24}$	0.31	13.357	1,4,7,-Cycloundecatri-ene, 1,5,9,9-tetramethy	13

TABLE 3.8 (Continued)

M.W.	Formula	Area %	R.T. min	Name	Peak
204	$C_{15}H_{24}$	0.05	13.850	1,3-Cyclohexadiene, 5-(1,5-dimethyl-4-hexe	14
204	$C_{15}H_{24}$	0.03	13.967	1,3,6,10-Dodeca-tetraene, 3,7,11-tri-methyl-,	15
206	$C_{12}H_{14}O_3$	14.85	14.138	3-Allyl-6-methoxyphe-nyl acetate	16
204	$C_{15}H_{24}$	0.04	14.217	Cyclohexene, 3-(1,5-dimethyl-4-hex-enyl)-6-	17
220	$C_{15}H_{24}O$	0.07	14.851	1H-Cycloprop[e]azulen-7-ol, deca-hydro-1,1	18
220	$C_{15}H_{24}O$	0.49	14.915	Caryophyllene oxide	19
222	$C_{15}H_{26}O$	0.07	15.500	2-Naphthalenemetha-nol, 1,2,3,4,4a,5,6,7-oc	20
220	$C_{15}H_{24}O$	0.12	15.550	Tetracy-clo[6.3.2.0(2,5).0(1,8)]tridecan-9-ol,	21
222	$C_{15}H_{26}O$	0.17	15.757	2-Naphthalenemetha-nol, decahydro-α	22
204	$C_{15}H_{24}$	0.05	16.617	Spiro[5.5]undec-2-ene, 3,7,7-trimethyl-11-m	23
212	$C_{14}H_{12}O_2$	0.07	17.018	Benzyl Benzoate	24
390	$C_{24}H_{38}O_4$	0.05	23.900	1,2-Benzenedicarbox-ylic acid, diisooctyl est	25

3.4 SUMMARY

Water distillation and steam distillation methods are common traditional methods that are used to distillate the essential oils. The novel technology for distillation of essential oils is ohmic heating, which can be used as an assisted water distillation. This technology reduces distillation time and it gives essential oils with high quality and high oil yield. On the other hand, steam distiller was been developed to reduce distillation temperature, by the author. In this case, the quality of essential oil was better than the traditional methods.

KEYWORDS

- Acids
- Acyclic monoterpene hydrocarbons
- Alcohol
- Aldehydes
- Alternative electric current
- Alternative voltage
- Ambient temperature
- Applied voltage
- Aromatic plant
- Aromatic plant membranes
- Blanching
- Boiling water
- Cell membrane
- Chemical composition
- Chemical compounds
- Chemical reaction
- *Cinnamomum camphora*
- Clove oil
- Clove oil yield
- Conductivity
- Cumin oil
- Decomposition
- Density of oil
- Diffusion
- Diffusion mechanism
- Disease control
- Diseases treatment
- Dissolution rate
- Distillation mechanism
- Distillation methods
- Distillation temperature
- Distillation time
- Distilled water
- Distilling flask
- Dying
- Dynamic energy
- Electric characteristics
- Electric conductivity
- Electric field intensity
- Electrical resistance
- Electrodes
- Electromagnetic
- Electromagnetic energy
- Electromagnetic spectrum
- Electromagnetic waves
- Energy consumption
- Energy efficiency
- Essential oil distillation
- Essential oil yield
- Essential oils
- Essential oils distiller
- Esters
- Etheric oils
- Eucalyptol
- Eucalyptol content
- Eucalypts leaves
- Eucalyptus oil
- Extraction temperature
- Extraction time
- Food
- Food industries
- Food ingredients
- Fouling
- Gelling agents
- Glass condenser
- Heat
- Heat energy
- Heat generates
- Heat transfer
- Heat transfer coefficient

- Heating rate
- Herbal products
- Homologue heating
- Hot water
- Hydro distillation
- Hydrocarbons
- Hydro-diffusion
- Hydro-distillation technique
- Hydrolysis
- Hydrolytic reactions
- Inner cylinder
- Ions concentration
- Joule heating
- Layer of hydrosol
- *Lippia alba*
- Mass transfer
- Mathematical modeling
- Medicinal
- Microwave
- Microwave assisted water distillation
- Microwave energy
- Milk pasteurization
- Monoterpene
- Novel technology
- Novel thermal treatment
- Ohmic heating
- Ohmic heating process
- Ohmic-hydro-distillation
- Oil glands
- Oil ingredients
- Oil quality
- Oil yield
- Oils extraction methods
- Osmosis
- Particles temperature
- Pharmaceutical
- Physical properties
- Physiochemical process
- Plants cell membranes
- Polymerization
- Proteins denaturation
- Reflective index
- Rosemary
- Second order mechanism
- Solid-liquid food
- Solid-liquid food mixture
- Specific gravity
- Steam
- Steam distillation
- Steam distillation method
- System performance coefficient
- Temperature
- Thermal characteristics
- Thermal treatment
- Traditional extraction method
- Vaporized oil
- Verdant technology
- Viscosity
- Viscosity of oil
- Volatile
- Volatile essence
- Volatile oils
- Voltage
- Voltage gradient
- Voltage regulator
- Volumetric heating
- Water
- Water distillation
- Water distillation method
- Whey proteins
- Whey proteins denaturation

REFERENCES

1. Abdull-Neby, A. A., (2009). Extraction and identification of essential oils from some plants and use in food systems and study their microbial inhibition. Thesis, Department of Food Science, University of Basrah, Basra, Iraq, pp. 100.
2. Abo-zaid, A. N., (1992). Aromatic plants and their products agricultural and medicine. 2nd Ed., Aldar Alarbia Press. Egypt, pp. 311.
3. Al-Mayah, A. A. A., (2001). Medicinal plants and treatment by herbals. Eiady Center for Studying and Publishing. Yemen, pp. 230.
4. Alhakeem, S. H., & Hassan, A. M., (1985). Food Processing. 1st Ed., Baghdad University Press, pp. 498.
5. Al-Hilphy, A. R. S., (2014). Practical study for new design of essential oils extraction apparatus using ohmic heating. *International Journal of Agricultural Science*, 4(12): 351–366.
6. Al-Hilphy, A. R. S., Al-fekaiki, D. F., & Hussein, R. A., (2015). Extraction of essential oils from some types of umbelifera family using microwave-assisted water distillation. *Journal of Biology, Agriculture and Healthcare*, 5(22), 16–28.
7. Al-Hilphy, A. R. S., (2015). Development of steam essential oils extractor. *IOSR Journal of Agriculture and Veterinary Science*, 8(12), 2319–2372.
8. Azar, Aberroomand P., Porgham-Daryasari, A., Saber-Tehrani, M., & Soleimani, M., (2011). Microwave-assisted hydro distillation of essential oil from *Thymus vulgaris* L. *Asian Journal of Chemistry*, 23(5), 2162–2164.
9. Baser, K. H. C., (1999). Industrial utilization of medicinal and aromatic Plants. *Proceedings of Second World Congress on Medicinal and Aromatic Plants for Human Welfare*, pp. 10–15.
10. Chemat, S., Ait-Amar, H., Lagha, A., & Esveld, D. C., (2005). Microwave-assisted extraction kinetics of terpenes from caraway seeds. *Chemical Engineering and Processing*, 44, 1320–1326.
11. Damayanti, A., & Setyawan, E., (2012). Essential oil extraction of fennel seed (Foeniculum vulgare) using steam distillation. *International Journal of Engineering Science*, 3(2), 12–14.
12. Eskilsson C. S., & Björklund, E., (2000). Analytical-scale Microwave-assisted Extraction. *Journal of Chromatography A, 902*, 227–250.
13. Fadel, O., Ghazi, Z., Mouni, L., Benchat, N., Ramdani, Amhamdi, M., Wathelet, H. J. P., et al., (2011). Comparison of microwave-assisted hydro distillation and traditional hydro ditillation methods for the Rosmarinus eriocalyx essential oils from Eastern Morocco. *Journal of Materials and Environmental Science*, 2(2), 112–117.
14. Gavahian, M., Farahnaky, A., Javidnia, K., & Majzoobi, M., (2012). Comparison of ohmic assisted hydro-distillation with traditional hydro-distillation for the extraction of essential oils from Thymus vulgaris L. *Innovative Food Science and Emerging Technologies, 14*, 85–91.
15. Geankoplis, C. J., (1993). Transport Processes and Unit Operation. 3rd Ed., Singapore. Prentice-Hill, pp. 280.
16. GNU. *GNU Free Documentation License*, Version 1.2 FL: Wikimedia, 2005.

17. Goullieaux, A., & Pain, J. P., (2014). Recent Development in Microwave Heating. In: *Sun, D. Ed., Emerging Technologies for Food Processing.* Academic Press, UK, pp. 361–377.
18. Goullieux, A., & Pain, J. P., (2005). Emerging technologies for food processing. In: Sun, D. W. Ed., *Ohmic heating.* Academic Press, Califonia, US, pp. 280.
19. Guenther, E., (1987). Essential Oils. Universitas Indonesia Press, Jakarta, pp. 695–709.
20. Handa, S. S., Khanuja, S., Longo, G., & Rakesh, D. D., (2008). Extraction Technologies for Medicinal and Aromatic Plants. Padriciano 99, 34012 Trieste, Italy, pp. 110.
21. Icier, F., & Ilicali, C., (2005). The effect of concentration on electrical conductivity of orange juice concentrates during ohmic heating. *European Food Research and Technology, 220,* 406–414.
22. Icier, F., Yildiz, H., & Baysal, T., (2008). Polyphenoloxidase deactivation kinetics during ohmic heating of grape juice. *Journal of Food Engineering, 85,* 410–417.
23. Kapás, A., András, C. D., Dobre, T. G., Vass, E., Székely, G., & Stroescu, M., et al., (2011). The kinetic of essential oil separation from fennel by microwave assisted hydro-distillation (MWHD), UPB Scientific Bulletin, Series B: *Chemistry and Materials Science, 73,* 113–120.
24. Karakaya, S., Nehir, S. E., Karagozlu, N., Sahin, S., Sumnu, G., & Bayramoglu, B., (2014). Microwave assisted hydro-distillation of essential Oil from rosemary. *Journal of Food Science and Technology, 51*(6), 1056–1065.
25. Lawrence, B. M., (1995). The isolation of aromatic materials from natural plant products. In: *Desiva, K. T. Ed.,* Amanualon the essential oils industry. UNIDO, Vienna, Austria, pp. 57–154.
26. Malle. B., & Schmickl, H., (2005). Therische leselbstherstellen. verlagdie werkstatt Gmb H Lotzestraße 24 a. D-37083 Gttingen. 23–27.
27. Milojević, S. Ž., Radosavljević, D. B., Pavićević, V. P., Pejanović, S., & Veljković, V. B., (2013). Modeling the kinetic of essential oil hydro-distillation from plant materials. *Hemijska industrija, 67*(5), 843–859.
28. Munir, A., & Hensel, O., (2009). Biomass energy utilization in solar distillation system for essential oils extraction from herbs. International conference for Biophysical and Socio-economic Frame Conditions for the Sustainable Management of Natural Resources (Tropentag 2009) on October 6-8 in University of Hamburg, Germany, pp. 54.
29. Öztekin, S., & Martinov, M., (2007). Medicinal and Aromatic Crops, Harvesting, Drying and Processing, Haworth Food and Agricultural Products Press TM, An Imprint of the Haworth Press, Inc., 10 Alice Street, Binhamton, New York 13904–1580 USA,
30. Pronpunyapat, J. Pakamas, C., & Chakrit. T., (2011). Mathematical modeling for extraction of essential oil from Aquilaria crassna by hydro-distillation and quality of agarawood oil. *Bangladesh Journal of Pharmacology, 6,* 18–24.
31. Rakotondramasy-Rabesiaka, L., Havet, J. L., Port, C., & Fauduet, H., (2008). Solid-liquid extraction of protopine from fumaria officinalis L. analysis determination, kinetic reaction and model building. *Separation and Purification Technology, 59*(2), 253–261.

32. Ranjitha, J., & Vijiyalakshmi, S., (2014). Facile methods for the extraction of essential oil from the plant species A. *International Journal of Pharmaceutical Sciences And Research, 5*(4), 1107–1115.

33. Resan, R. A., (2016). Study of effect extraction methods on properties of essential oils for some types of Umbellifera family using GC MS. MSc Thesis, Department of Food Science, University of Basrah, Basra, Iraq, pp. 130.

34. Routray, W., & Orsat, V., (2012). Microwave-assisted extraction of flavonoids: A review. *Food and Bioprocess Technology February, 5*(2), 409–424.

35. Runha, F. P., Cordeiro, D. S., Pereira, C. A. M., Vilegas, J., & Olivera, W. P., (2001). Production of dry extracts of medicinal brazilian plants by spouted bed process: development of the process and evaluation of thermal degradation during the drying operation *Food and Bioproducts Processing, 79*(3), 160–168.

36. Saini, N., Singh, G. K., & Nagori, B. P., (2014). Physicochemical characterization and spasmolytic activity of essential oil of cumin (*Cuminum cyminum* Linn.) from Rajasthan, *International Journal of Biology, Pharmacy and Allied Sciences, 3*(1), 78–87.

37. Sastry, S. K., (2005). Advances in ohmic heating and moderate electric field (MEF) processing. In: Cano, M. P., Maria, Tapia, S. and Barbosa-Canovas, G. V. Eds., *Novel Food Processing Technologies*. Marcel Dekker, CRC Press, pp. 491–499.

38. Şen, S., & Yalçın, M., (2010). Activity of commercial still waters from volatile oils production against wood decay fungi. *Maderas Ciencia y Tecnología, 12*(2), 127–133.

39. Shirsat, N., Lyng, J. G., Brunton, N. P., & McKenna, B., (2004). Ohmic processing: electrical conductivities of pork cuts. *Meat Science, 67*, 507–514.

40. Stashenko, E. E., Jaramillo, B. E., & Martinez, J. R., (2004). Comparison of different extraction methods for the analysis of volatile secondary metabolites of lippia alba *(Mill)* N.E. brown, grown Colombia and evaluation of its antioxidant activity. *Journal of Chromatography A, 1025*(1), 93–103.

41. Thostenson, E. T., & Chou, T. W., (1999). Microwave processing: fundamentals and applications. *Composites Part A: Applied Science and Manufacturing, 30*(9), 1055–1071.

42. Toledo, R. T., (2007). Fundamentals of Food Process Engineering. 3rd Ed., Springer Science + Business Media, LLC, pp. 579.

43. Wei, L., Zhang, Y., & Jiang, B., (2013). Comparison of microwave assisted hydro-distillation with the traditional hydro-distillation method in the extraction of essential oils from dwarfed *Cinnamom camphora* var. *Linaolifera Fujita* leaves and twigs. *Journal of Food Science and Technology, 5*(11), 1436–1442.

44. Zareifard, M. R., Ramaswamy, H. S. Trigui, M., & Marcotte, M., (2003). Ohmic heating behavior and electrical conductivity of two-phase food systems. *Innovative Food Science and Emerging Technologies, 4*, 45–55.

PART II

ENGINEERING INTERVENTIONS IN FOODS AND PLANTS FOR HEALTH BENEFITS

CHAPTER 4

PROCESSING TECHNOLOGY AND POTENTIAL HEALTH BENEFITS OF COFFEE

VISHAL SINGH and DEEPAK KUMAR VERMA

CONTENTS

4.1   INTRODUCTION

Coffee has been produced worldwide with total production of 150.4 million bags (each bag contains 60 kg) and India has produced 5.2 million bags of Arabica and 3.5 million bags of Robusta in 2013. The coffee fruit consists of pericarp (smooth and hard outer skin), mesocarp (soft yellowish, fibrous pulp) layered with followed by pectin layer (layer of thin, transparent and colorless mucilage) and then layer of endocarp (yellowish in color) followed by silver skin covers the coffee bean individually [15]. Coffee is very popular worldwide [61]. Many tropical and subtropical countries produce coffee as a main agriculture product [35]. For more than 500 years, coffee has been used as drink and now-a-days it is one of the most common used beverages. The first coffee beans toasting and preparation of drink from it were observed in Persia in 16th century.

Coffee have more than one hundred species out of which, Arabica and Robusta species of coffee has been contributing three fourth and one fourth of total production, respectively [90, 118]. Most of all the species of coffee differ from each other due to their chemical composition [26]. Coffee has several varieties and characteristics that depend on climatic/soil factors, its flavor, acidic content, texture and aroma, etc. [67, 78, 117]. Robusta variety of coffee contains more resistance towards diseases than Arabica [15, 38]. Arabica species of coffee originated from Ethiopia [6, 49]. Enhancement in the quality through good cultivation and processing practices and proper channel of marketing can enlarge the coffee trade [68] because market performance is mainly based on consumer acceptance, which is directly proportional to the quality of product [19].

FIGURE 4.1   Coffee Plant Loaded with Flowers, and Cherries.

Coffee plant (Figure 4.1) carries the fruits (in raw condition it is called cherry), which is harvested after ripening or when its color change to red [11]. As demand of coffee increases, processing and value additions have also been accelerated. Today, two different process known as dry and wet methods are in practice for removing the pulp and hull [81]. Arabica species contains lesser caffeine and more robust against insect and pest attacks in comparison with Robusta variety of coffee. For healthy and good productivity of Arabica variety, soil of 5.2 to 6.3 pH range is more suitable [118]. Amount of produced microbial volatile compounds during fermentation causes good aroma of coffee [55, 56]. Quality improvement of coffee can take place by focusing on parameters like shape, size, color and hardness as well as odor and textural properties [43, 47, 51, 58].

Coffee is a popular and worldwide consumable drink prepared by extracting of soluble elements of roasted coffee beans powder in the boiling water. Flavor, aroma and composition of coffee beverages altered with time and temperature of roasting and extraction in water. Coffee contains different sugars (e.g., sucrose, glucose, fructose and mannose, etc.) and amino acids like alanine, asparagines, glutamic acid and lysine, etc. [15, 57] besides of vitamins like B_3 and B_{12} [15, 69, 110, 111]. In addition, coffee also contains some other very useful chemical constituents like antioxidants, fiber and melanoidins, etc. [17]. The chemical composition of coffee has been influenced with the combination of time and temperature [31, 33, 92] besides processing technologies [21, 63] and duration of storage [86]. Brewed coffee from typical ground coffee prepared with tap water contains 40 mg caffeine per 100 g and no essential nutrients in significant content [106]. In espresso, however, likely due to its higher amount of suspended solids, there are significant contents of magnesium, the B-vitamins, niacin and riboflavin, and 212 mg of caffeine per 100 g of ground coffee [https://en.wikipedia.org/wiki/Coffee].

Usefulness of freshly processed coffee is more because of its higher viability [60] besides of its active metabolism [95, 96]. Hydrolysis and oxidation also affect the different compounds present in coffee [66]. Due to roasting of coffee, vanillin content were increased [31] as well as physical properties of coffee (e.g., color, texture, density and different volatile compounds, etc.) has also influenced [58]. To obtain an adequate physical appearance and amount of volatile compounds, the roasting of coffee should be performed carefully because it is a tedious operation of a coffee processing [48]. For preserve the quality, coffee should be cooled down just after roasting [14, 37]. Super-heated water up to 85–90°C is used for coffee extraction and steam with 7 to 9 bars pressure applied for 25 to 35 seconds to ramp up the filtration rate [85]. Fresh and roasted (roasted till dark brown color), coffee powder is used to prepare strong and cloudy drink [15, 76]. Chemical constituents of coffee brews depend upon the different factors of processing like water temperature and pressure during its extraction, grinding time and volume of coffee, etc. [3–5, 16, 22, 50, 63, 75, 82, 88]. Soluble coffee contributes nearly half of the total coffee production of the world [87, 109].

The coffee is the most worldwide accepted beverage and is considered of commercial value. Due to its vast acceptability as a drink, number of producing and processing industries has existed throughout the world which produce huge amount of liquid and solid wastes [74]. Cunha [29] stated that nearly 77 million bags of coffee had been dumped into the low land area and into the ocean during third and fourth decades of the 20th century causing severe pollution that has not drawn significant attention to cope with waste management. However, in the 21st century, waste management had been considered as a intensive topic of research and several technologies have come to use the residues of coffee industries as a parent material for other industries, recycling and refining it for getting finer products, etc. [73, 99].

This chapter reviews processing technologies and potential health benefits of coffee.

FIGURE 4.2 Green and Roasted Coffee Beans: (A) Arabica (*Coffea arabica*), (B) Robusta (*Coffea canephora*), and (C) Liberica (*Coffea liberica*).

4.2 TYPES OF COFFEE

The international coffee trade is made almost exclusively on green (raw) beans (Figure 2). In this form, Arabica and Robusta beans are easily distinguished by their physical characteristics (e.g., size, shape, and color) [2]. Rubiaceae family contains more than 70 species in which mainly two varieties (Arabica and Robusta) are commercially grown at large level.

4.2.1 ARABICA (COFFEA ARABICA)

An Arabica variety has small plant of 5 m in height. Arabica is grown in tropical regions at an altitude of over 500 m, but preferably at 1000–1500 m. Higher altitudes generally produce a better quality crop [65]. Arabica has a milder and more flavorful taste and lower caffeine content than Robusta, which is more resistant to insect damage and disease [118]. Farah [39] observed that flavor of Arabica coffee is more effective when beans are slightly roasted; and its market price is also higher than Robusta due to its flavor and other qualitative subjects.

4.2.2 ROBUSTA (COFFEA CANEPHORA)

Robusta variety is stronger and resistant against diseases caused by insects and pests. It is used to produce instant coffee and for enhancing the body and the foam of some coffee brews [102]. The shape of Cherries and beans of Robusta is more round than the Arabica species and Robusta can grow up to height of 7–10 m when fully mature [2]. It contains more amount of caffeine, antioxidants and soluble solids and presence of acids like chlorogenic acids that protect the plants from radiation, insect-pests and microorganism and making coffee plants more robust [39, 40].

4.2.3 LIBERICA (COFFEA LIBERICA)

Liberica is a larger tree than above two species and it can attain a maximum height of 18 m with large berries. The coffee brew of Liberica beans is very bitter and sour and yields are also very low. Harvesting of coffee from tall trees is a tedious job. The low quantity and quality of the crop has reduced the amount of Liberica [2].

4.3 DIFFERENT COFFEE PRODUCTS

4.3.1 GREEN COFFEE

Scientific practices from plantation to consumption are highly essential for obtaining fine quality of green coffee (Figure 4.3A) with good nutrition value because unhygienic harvesting, handling and storage practices can hamper the qualitative aspects like texture, flavor, color, nutrients and bioactive compounds [23, 44, 106]. Wet or dry method of pulp extraction

FIGURE 4.3   Different coffee products: (A) Green coffee, (B) Decaffeinated coffee, (C) Steam-treated coffee, (D) Monsooned coffees, (E) Roasted coffee, (F) Instant coffee, and (G) Coffee beverage (Brew).

is a operation after harvesting of coffee cherries. In dry method, coffee beans are exposed to sun for drying (after final, moisture content should be nearly 10–12%), just after pulp removal and meanwhile damaged, discolor and immature beans should be separated [39, 108] and stored as green coffee bean for consumption. Green coffee contains different compounds like fructose, glucose, mannose, arabinose, and rhamnose, raffinose and stachyose, etc. [64].

4.3.2   DECAFFEINATED COFFEE

Decaffeinated coffee (Figure 4.3B) fulfills the demand of consumers who have health problems like insomnia, anxiety, nervousness and stress or any other health disorder [97]. Market of decaffeinated coffee is increasing worldwide due to increasing the population of health conscious of persons [70, 79, 100]. Caffeine removed from coffee by different extraction method and organic solvents like ethyl acetate or water has been used for cleaning the beans before and after extraction. Coffee beans should be dried after decaffeination to an optimum moisture content as it was before extraction of caffeine [40, 42]. It has been utilized to prepare different consumable products like chocolates, cold drinks, etc. besides medicine preparation. Decaffeinated coffee can generally loose its original flavor [39, 40].

4.3.3   STEAM-TREATED COFFEE

Steamed coffee (Figure 4.3C) is prepared by steaming the coffee beans just before roasting. Due to steaming, chlorogenic acid content is reduced. This coffee is preferred by consumers with stomach problem [39].

4.3.4   MONSOONED COFFEE

Monsooned coffee (Figure 4.3D) is prepared with sorted out and healthy beans. In Malabar region of southern India, during monsoon season, beans should be exposed to wind for 90 to 120 days for natural removal of moisture. Its good demand in different region of the world is due to unique flavor, aroma and chemical composition [39, 113].

4.3.5 ROASTED COFFEE

Roasted coffee (Figure 4.3E) includes some properties in the coffee beverage like flavor and aroma through some reactions, which can alter the chemical composition [20]. Roasting of coffee affect the level of anti-oxidant activity and it has been noticed that medium roasted coffee has a maximum oxidant activity because of most favorable condition for decomposition of phenolic compound [36]. The flavor and aroma of coffee are due to formation of some volatile compounds during roasting process [15]. For maintaining the good textural appearance and flavor besides availability of chemical compounds in coffee beans, temperature should be controlled during roasting because due to variations in roasting temperature and time, characteristics of different compounds like protein and sugar, etc. can also be altered [25].

4.3.6 INSTANT COFFEE

Instant coffee (Figure 4.3F) is prepared by water-soluble compounds present in roasted coffee. For producing instant coffee, good quality roasted coffee is boiled with water at high temperature and pressure for obtaining the water-soluble compounds. After getting the water-soluble compounds, it should be cooled and dried up to nearly 5% moisture content. Collect and store this dried powder as instant coffee [1, 24]. Different types of dryers like freeze dryer; spray dryer and solar dryer, etc. can be used for drying of instant coffee. Ground roasted healthy beans of Robusta produce more amount of instant coffee because of its high content of soluble solids [39].

4.3.7 COFFEE BEVERAGE (BREW)

Preparation of coffee brew (Figure 4.3G) involves different techniques like pouring, dripping or spraying roasted coffee beans with hot water spraying, and finally filtering operation is performed. The method employed (like boiling, pouring, dripping or spraying with hot water) for coffee brewing has a direct influence on the different constituent present in coffee beans [103]. During brewing, water of temperature and pressure is 90–95°C and 9 atmospheres. Due to this process, water-soluble compounds and other extracts are collected, which are known as coffee brew [39].

4.4 COFFEE PROCESSING TECHNOLOGY

Coffee processing involves different operations like roasting, grinding and packing, which is done with the help of different machines (Figures 4.4–4.6). Hygienic and systematic operations are necessary for good quality of coffee. Dry and wet methods are two important methods for coffee processing. Generally, sun drying of coffee beans is used in case of dry method. This method is quicker than wet method and is mostly practiced in African and Asian regions; and in southern parts of the United States, wet method is practiced. Beans are separated from coffee fruits after plucking from plants and then undergo for fermentation. After completion of fermentation, these are washed and dried to optimum moisture. Generally ripen coffee fruits are suitable for de-pulping because of their soft skin. Coffee beans should be kept for drying just after washing the de-pulped beans.

Quality of coffee is often based on the aroma, flavor and color, etc. However, due to findings of several researches on the availability of nearly 800 different chemicals and their beneficial characteristics, the importance of quality has increased due to worldwide acceptability of coffee [42]. Appropriate and scientific processing methods and its approaches are essential for retaining of chemical compounds [22, 63]. Different researchers have observed that quality of coffee also depends on its variety, climate of cultivation, roasting process and its methods [11, 22, 98, 112]. Quality of coffee processed with wet method is generally better than dry processing [96]. Stress metabolism takes place due to drying that influences the chemical compounds content available in coffee [21]. Moisture less than 12% is adequate for hindering the microbial growth besides the formation of mycotoxin [89].

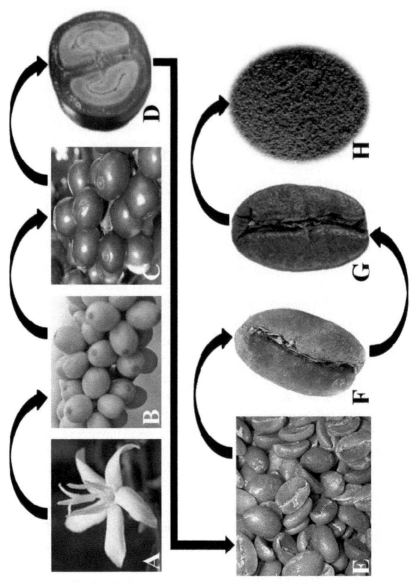

FIGURE 4.4 Stepwise growth and processing of coffee beans from flowering stage to coffee powder: (A) Flower of coffee plant, (B) Raw cherry (green cherry), (C) Ripen cherry (red cherry), (D) Cross section of ripen cherry, (E) Bean with sliver skin, (F) Dried bean, (G) Roasted bean (red-brown bean), and (H) Coffee powder.

Intensive research reveals that nearly 800 chemical elements have been noticed in roasted coffee which strongly influences the flavor and other characteristics of coffee [40, 42]. Duration of storage affect the quality of coffee. If storage duration is less then qualitative attributes should be better [12, 28, 60, 94, 119]. Lipids content of coffee seed has been influenced due to oxidation during storage [62, 103].

Quality of coffee seed is directly proportional to its viability [101]. In general, coffee seeds can be stored for less duration, thus for storing the seed for long durations, favorable storage conditions are required

FIGURE 4.5 Flow chart for coffee processing.

FIGURE 4.6   Different machines for coffee processing: A) Packing machinery; B) Polisher; C) Pulper; D) Destone; E) Coffee grinder; F) Coffee bean separator; G) Coffee roaster; H) Drum dryer.

[91, 116]. Carbohydrates and free amino acids content of coffee beans is degraded during roasting [45]. Green coffee contains sucrose as a main carbohydrate [18, 63]. Due to decaffeination, components responsible for good flavor are also reduced [100]. Degree of roasting influences the color of coffee beans [25]. Coffee seed also contains chemical compounds like theophylline and theobromine [70]. Decaffeinated coffee has lesser antibacterial effect against cariogenic bacteria than the caffeinated one [8].

Several researchers have noticed that coffee has good antioxidant property due to presence of chlorogenic acids [52, 83, 93, 105]. Flament et al. [46] and Kolling-Speer et al. [64] stated that green coffee contains different carbohydrates (e.g., poly and oligosaccharides). Monsooned coffee processing is practiced in some parts of India like Kerala, Karnataka and some other parts of Malabar Coast where coffee seeds are exposed in the monsoon rain which enhances the quality [113]. Coffee contains more alcohols in stage of ripening than seed [80], and moisture content of roasted coffee ranges from 1.5 to 5% [110].

Researchers reveal that melanoidins, which forms during Millard reaction, have important role in antioxidant and antibacterial characteristics of coffee [32, 33, 77]. Reduction of mono and polysaccharides contents can be responsible for increasing the acidic nature of coffee brew [24, 54]. Different carbon, oxygen, sulphur and nitrogen containing volatile compounds have been noticed in roasted coffee [7, 107]. Figure 4.7 indicates how color changes due to caramelization of sugar and swelling of beans was started at about 130°C during roasting and beans turned in brown color at ≥160°C [39].

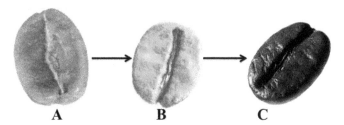

A   **B**   **C**

FIGURE 4.7   Color change due to caramelization with temperature and time of roasting: (A) Fresh and dried bean, (B) Medium degree of roasting at low temperature or less roasting duration, and (C) High degree of roasting at high temperature or more roasting duration.

4.4.1 PICKING

Only pink red or yellow cherries are harvested every week or every two weeks, depending on the size of coffee beans. Coffee can be harvested throughout the year, but the harvesting peak season is from August to September and February to March. Picking days differ in each region, depending on the marketing demand [104]. After picking, most farmers directly pulp the cherries, using traditional pulping machines (wooden type pulper) in their own plantations. Only matured coffee beans (which can be examined by squeezing the coffee fruits gently) are picked.

4.4.2 FLOATING

It is recommended to float the cherries before pulping, other farmers may not. The purpose of the floating is to separate good cherries and bad cherries, a kind of sorting technique. Bad cherries will be floated and good cherries will be sunk down. Good cherries are pulped with traditional machines. Due to this step of processing coffees from these regions are known for better quality (less defects and higher yields), hence fetch better price compared to coffees from other regions [104].

4.4.3 PULPING

Pulping is a process to separate outer skin from the bean. Mostly cherries are pulped directly after harvesting at the coffee plantation [104]. Pulping has been very difficult when cherries are hard and green. Generally on small-scale units, the cherries can be pulped with the help of pestle and mortar. Drum and disc pulpers have generally been used on small-scale industries [114]. Proper cleaning of pulper and pulper tanks is necessary before pulping operation due to hygienic point of view. Pulper operation unit is adjusted according to fruit size and proper precaution should be taken for reducing the cut or any other damage because high yield of deformed beans can reduce the quality of parchment [115]. Before pulping, all the undesired portions like leaves, twigs and stones are removed and only clean water should be used in processing and whenever sufficient cleaned water is not available then the water used in pulping process can be recycled for second run but never can be used on next day. For getting good quality of pulping, the fresh harvested coffee fruits have to be pulped on same day of harvesting. Skin of the pulp is removed and kept away from the place of pulping for preventing the contamination of parchment coffee.

4.4.4   FERMENTATION

Generally, fermentation techniques are done by dry and wet methods. Plastic bag is used for storing around 12–14 hours in dry fermentation process and on the other hand, wet fermentation is performed by soaking parchment in plastic pails up to 12–14 hours [104]. In the dry process, the ripe coffee beans (coffee cherries) are dried in the sun after which the skin is removed to produce the green beans. In the wet process, coffee cherries are pulped and fermented to get rid of the mucilage that adheres to the beans. After the fermentation, beans are washed and dried [72].

Natural fermentation gives better results than other method of fermentation. Fermentation process should be considered carefully because of it is one of the most effective steps of coffee processing which influence the flavor and aroma of coffee. Fermentation process in the case of Robusta coffee is completed within 36 to 72 hours because of its thick and hard mucilage. Over fermentation should not be permitted and in same way under-fermented coffee beans are not allowable because under-fermented beans create mustiness in the cup [115]. Acidity increases and sliver skin of the coffee beans is removed after fermentation [39].

4.4.5   WASHING

Coffee should be washed out to remove the mucilage and to separate remaining pulp from coffee parchment. During washing process, qualitative and damaged coffee are separated [104]. Washing machineries are cleaned properly before preparation of parchment coffee [115]. Coffee beans are completely washed before fermentation. Semi-washing operation also can be implemented in which ripe coffee beans are pulped and dried without removing the mucilage [72]. After washing coffee, bins are stirred by rubbing the coffee beans to each other for easy breaking of the coffee beans and removal of remaining mucilage [34].

4.4.6 DRYING

Different method are used for drying the coffee like solar drying (solar cabinet and solar dryer) and artificial drying [114]. For checking the drying status of coffee beans, one can observe the sound of handful beans, if sound is rattling means drying is over. In sun drying method, coffee should not be exposed during afternoon (because of high temperature during afternoon) especially third and fourth day of drying because during this period if beans will be exposed then chances of cracked beans can be increased. Proper stirring of coffee periodically facilitates the uniform drying and coffee is completely dried within 10–12 days. Coffee beans should be dried to 12–12.5% moisture level [115].

4.4.7 STORING

Usually, rural farmers pack the wet parchment in woven plastic bags and harvest their cherries three days before the marketing day. The first two days are allocated to process their cherries [104] but scientifically a well-designed and safe storage is essential for maintaining the optimal conditions like low temperature, low humidity and free from pests. The storage should be located in a shaded, dry place. The produce should be checked regularly and if it has absorbed too much moisture it should be dried again and to prevent the pests entering, the roof should be completely sealed [114]. Aroma of coffee bean can be reduced due to storage in atmospheric conditions because oxidation of roasted or grounded coffee will be increased due to more exposed surface which facilitates the availability of oxygen; that is why sealed pack is desired for packaging of coffee. If different process likes grinding, degassing have taken place within controlled atmosphere, then very high, protective and stable packaging is required for final product [13].

4.5 POTENTIAL HEALTH BENEFITS

Coffee have beneficial impact on human body like caffeine (known as a stimulant) and several chemical compounds has been noticed in this beverage to have beneficial properties for the human health. Coffee contains different compounds that may enhance the antioxidant properties and protect the cardiovascular system, nervous system, reproductive system, cholesterol level [53].

Coffee consumption is very beneficial for reduction of liver disease. Research has reported that daily consumption of 4 cups of coffee reflects 84% reduction in Cirrhosis and also reduces the chronic liver disease. Several scientists stated that coffee is a good source of caffeine which is responsible for prevention of cognitive decline. Daily consumption of 5 cups of coffee can reduce the Parkinson's by 60% [71]. Regular coffee consumption can reduce effect of cancer up to 13%. Normal metabolic process produces some oxidizing agent that causes DNA injury which also can be controlled by coffee consumption [71]. In USA, intensive health study based on 88,259 women of 26–46 year old indicated that the daily consumption of coffee reduced the risk of diabetes compared to non-coffee drinkers who showed 87% of relative risk of diabetes (in the case of 1 cup per day coffee consumption), 58% (in the case of 2–3 cup daily consumption) and 53% (in the case when consumer consume more than 4 cup). Regular consumption of coffee also showed that it also reduces the risk of liver injury. The consumers of coffee more than two cups in a day have reduced the 50% chances of liver disease in the comparison of non-drinkers [84]. Coffee also contains several micronutrients like magnesium, potassium, niacin and vitamin E, which provide adequate nourishment to humans. According the USDA research, 240 ml of brewed coffee provided 7 mg of magnesium. One cup of coffee provided 1–5% recommended dietary allowance of magnesium in adult men and one cup of coffee contributes only 1–2% of the adequate intake of potassium (4700 mg/d) in adults [59]. Several research studies have shown that coffee helps in glucose synthesis and release in the body which accomplish the amount of glucose by enhancing the pathway of glucose-6-phosphatase (glucose-regulating enzyme), which reduces the sugar levels in the blood. Research report showed 57% reduction in breast cancer in women who consumed 5 cups of coffee every day. Some other constituents like chlorogenic acid, caffeic acid and phytoestrogens have also been found in coffee, which play a vital role in maintaining a good health.

Different climatic factors like topography, soil characteristics, humidity, temperature and sunshine hours, etc. also affect the quantitative and qualitative characteristics of coffee seeds [9, 10, 21]. Anthocyanins and lignans which are available in very less quantity in green coffee beans assumed as residues of coffee cherry [41]. It is noticed that available quantity of theophylline and theobromine in the coffee beans has been beneficial in caffeine metabolism [70]. Adequate amount of caffeine causes bitter taste besides used as refreshment agent of coffee drink but excess amount of caffeine causes insomnia, headache and nervousness, etc. [27, 42, 79, 108]. Caffeine also reduces the effect of microorganisms like cariogenic (this microorganism is responsible for teeth decaying), etc. [8].

4.6 SUMMARY

Production and consumption of coffee is rapidly increasing throughout the world by which demand is also increasing day to day. Some products of coffee like roasted, instant, monsoon, decaffeinated, steam treated and brew coffee have some different characteristics, compostion besides the nutrients contents. All the variety of coffee products have good market, so for maintaining the demand of coffee, quality and other characteristics should be high. For enhancing the quality of coffee and its product, different processing technology has been involved. Coffees has a good impact on human health and eliminates different disease related to cardiovascular system, nervous system, reproductive system and also reduce the cholesterol level, DNA injury, diabetes, sugar levels, Chronic liver disease etc because of it contains carbohydrate, vitamins, antioxidants, enzymes and some other water-soluble active compounds. Several advanced technologies has been introduced for coffee processing, preservation and preparation by research scientist and industrial experts with aim of obtaining qualitative and hygienic coffee products.

Above studies focused the importance and significant role of coffee as a consumable drink which protect the human body from several diseases that is why deep and qualitative study has been necessary for in lighting all characteristics of coffee.

KEYWORDS

- Acidic content
- Acids
- Active metabolism
- Agriculture
- Agriculture product
- Alanine
- Alcohols
- Amino acids
- Anthocyanins
- Antibacterial effect
- Antioxidant activity
- Antioxidants
- Anxiety
- Arabica
- Arabica beans
- Arabica coffee
- Arabica species
- Arabica species of coffee
- Arabica variety
- Arabica variety of coffee
- Arabinose
- Aroma
- Aroma of coffee
- Artificial dryer
- Artificial drying
- Asparagines
- Bean
- Beverages

- Bioactive compounds
- Brewing
- Brown color
- Caffeine
- Caffeine content
- Caffeine metabolism
- Cancer
- Caramelization
- Carbohydrates
- Cardiovascular system
- Cariogenic
- Cariogenic bacteria
- Chemical composition
- Chemical composition of coffee
- Cherry
- Chlorogenic acid
- Cholesterol level
- Chronic liver disease
- Cirrhosis
- Cloudy drink
- Coffea Arabica
- Coffea canephora
- Coffea liberica
- Coffee
- Coffee bean
- Coffee bean separator
- Coffee beverage
- Coffee brew

- Coffee extraction
- Coffee fruit
- Coffee grinder
- Coffee industries
- Coffee plant
- Coffee plantation
- Coffee powder
- Coffee processing
- Coffee production
- Coffee roaster
- Coffee seed
- Coffee trade
- Cognitive decline
- Color
- Cultivation
- Decaffeinated coffee
- Degree of roasting
- Density
- De-pulped beans
- De-pulping
- Diabetes
- DNA injury
- Drink
- Drum dryer
- Dry and wet fermentation
- Dry and wet methods
- Dry fermentation
- Dryer
- Drying
- Drying of coffee
- Endocarp
- Ethyl acetate
- Extraction method
- Extraction of caffeine
- Fermentation
- Fermentation process
- Fermented beans
- Fiber
- Filtration rate
- Flavor
- Floating
- Food
- Food commodity
- Free amino acids
- Freeze dryer
- Fructose
- Glucose
- Glucose synthesis
- Glucose-6-phosphatase
- Glucose-regulating enzyme
- Glutamic acid
- Green (raw) beans
- Green coffee
- Green coffee bean
- Grinding
- Grinding time
- Ground roasted healthy beans
- Health
- Health benefits
- Health disorder
- Health problems
- Human health
- Hydrolysis
- Insomnia
- Instant coffee
- Liberica
- Liberica beans
- Lignans
- Liver disease
- Liver injury
- Lysine
- Mannose
- Market price
- Melanoidins
- Mesocarp
- Microbial

REFERENCES

1. Adams, M. R., & Dougan, J., (1987). *Waste Products*. In: Clarke, R. J., Macrae, R. Eds. Coffee, volume 2, New York: Elsevier Science, pp. 533.
2. Alves, R. C., Casal, S., Alves, M. R., & Oliveira, M. B., (2009). Discrimination between Arabica and Robusta coffee species on the basis of their tocopherol profiles. *Food Chemistry, 114*(1), 295–299.
3. Andueza, S., Maeztu, L., Dean, B., de Peña, M. P., Bello, J., & Cid, C., (2002). Influence of water pressure on the final quality of Arabica espresso coffee. Application of multivariate analysis. *Journal of Agricultural and Food Chemistry, 50*, 7426–7431.
4. Andueza, S., Maeztu, L., Pascual, L., Ibanez, C., de Peña, M. P., & Cid, C., (2003). Influence of extraction temperature on the final quality of espresso coffee. *Journal of the Science of Food and Agriculture, 83*, 240–248.
5. Andueza, S., Vila, M. A., Peña, M. P., & Cid, C., (2007). Influence of coffee/water ratio on the final quality of espresso coffee. *Journal of the Science of Food and Agriculture, 87*, 586–592.
6. Anthony, F., Bertrand, B., Quiros, O., Lashermes, P., Berthaud, J., & Charrier, A., (2001). Genetic diversity of wild coffee (*Coffea arabica L.*) using molecular markers. *Euphytica, 118*, 53–65.
7. Antonio, A. G., Iorio, N. L. P., Pierro, V. S. S., Candreva, M. S., Farah, A., dos Santos K. R. N., et al., (2011). Inhibitory properties of coffea canephora extract against oral bacteria and its effect on demineralisation of deciduous teeth. *Archives of Oral Biology, 56*(6), 556–564.
8. Antonio, A. G., Moraes, R. S., Perrone, D., Maia, L. C., Santos, K. R. N., Iorio, N. L. P., et al., (2010). Species, roasting degree and decaffeination influence the antibacterial activity of coffee against streptococcus mutans. *Food Chemistry, 118*, 782–788.
9. Arnold, U., & Ludwig, E., (1996). Analysis of free amino acids in green coffee beans, II. Changes of the amino acid content in *Arabica coffees* in connection with postharvest model treatment. *Zeitschrift für Lebensmittel-Untersuchung und-Forschung, 203*, 376–384.
10. Arnold, U., Ludwig, E., Kuhn, R., & Moschwitzer, U., (1994). Analysis of free amino acids in green coffee beans, I. Determination of amino acids after precolumn derivatization using 9-fluoromethylchloroformate. *Zeitschrift Für Lebensmittel-Untersuchung und-Forschung, 199*(1), 22–25.
11. Arya, M., & Rao, J. M., (2007). An impression of coffee carbohydrates. *Critical Reviews in Food Science and Nutrition, 47*, 51–67.
12. Bacchi, O., (1958). Studies on the seed storage (*Estudos sobre a conservação de sementes*), IV. Café. *Bragantia, 17*, 261–270.
13. Baggenstoss, J., Poisson, L., Kaegi, R., Perren, R., & Escher, F., (2008). Coffee roasting and aroma formation: application of different time-temperature conditions. *Journal of Agricultural and Food Chemistry, 56*(14), 5836–5846.
14. Baggenstoss, J., Poisson, L., Luethi, R., Perren, R., & Escher, F., (2007). Influence of water quenches cooling on degassing and aroma stability of roasted coffee. *Journal of Agriculture and Food Chemistry, 55*, 6685–6691.

15. Belitz, H. D., Grosch, W., & Schieberle, P., (2009). Coffee, tea, cocoa. In: Belitz, H. D., Grosch, W., Schieberle, P. Eds. Food Chemistry. 4th Ed., Leipzig: Springer, pp. 938–951.
16. Bell, L. N., Wetzel, C. R., & Grand, A. N., (1996). Caffeine content in coffee as influenced by grinding and brewing techniques. *Food Research International, 29*, 185–189.
17. Borrelli, R. C., Esposito, F., Napolitano, A., Ritieni, A., & Fogliano, V., (2004). Characterization of a new potential functional ingredient: coffee silverskin. *Journal of Agricultural and Food Chemistry, 52*, 1338–1343.
18. Bradbury, A. G. W, (2001). Carbohydrates. In: Clarke, R. J., Vitzthum, O. G. Eds., Coffee. Recent Developments. Oxford: Blackwell Science, pp. 1–17.
19. Bruhn, C. M, (2002). Consumers issues in quality and safety. In. kadar, a. a. ed., postharvest technology of horticultural crops. Agriculture and Natural Resources, University of California, California, pp. 31–37.
20. Buffo, R. A., & Cardelli-Freire, C., (2004). Coffee flavor: an overview. *Flavor and Fragrance Journal, 19*, 99–104.
21. Bytof, G., Knopp, S. E., Schieberle, P., Teutsch, I., & Selmar, D., (2005). Influence of processing on the generation of g-aminobutyric acid in green coffee beans, *European Food Research and Technology, 220*, 245–250.
22. Bytof, G., Selmar, D., & Schieberle, P., (2000). New aspects of coffee processing: How do the different post-harvest treatments influence the formation of potential flavor precursors, *Journal of Applied Botany, 74*(3–4), 131–136.
23. Cirilo, M. P. G., Coelho, A. F. S., Araujo, C. M., Goncalves, F. R. B., Nogueira, F. D., & Gloria M. B. A., (2003). Profile and levels of bioactive amines in green and roasted coffee., *Food Chemistry, 82*, 397–402.
24. Clarke, R., (1985). *Coffee Chemistry.* Clarke, R. J., Macrae, R. (Eds.). 1st Ed., Vol.-1, Elsevier Applied Science Publishers, UK, pp. 115.
25. Clarke, R. J., (2003). Coffee: Green coffee/roast and Ground. In: Caballero, B.; Trugo, L. C.; Finglas, P. Eds., Encyclopedia of Food Science and Nutrition, volume 3. 2nd Ed. Oxford: Academic Press, pp. 120–125.
26. Clifford, M. N., (1985). Chemical and physical aspects of green coffee and coffee products. In: clifford, M. N., Willson, K. C. Eds., Coffee Botany, Biochemistry and Production of Beans and Beverage. Croom Helm, London, pp. 305–374.
27. Clifford, M. N., (2000). Chlorogenic acids and other cinnamates nature, occurrence, dietary burden, absorption and metabolism. *Journal of the Science of Food and Agriculture, 80*, 1033–1043.
28. Couturon, E., (1980). Maintaining seed viability of coffee by controlling the storage temperature (*Le maintien de la viabilité des graines de caféiers par le contrôle de la température de stockage*).*Café, Cacao, 24*, 27–32.
29. Cunha, M. R., (1992). Statistics Appendices (*Apêndice estatístico*). In: Bacha E. L.; Greenhill, R. Eds., 150 years of Coffee in Rio de Janeiro: Marcellino Martins & E. Johnston, pp. 286–388.
30. Czerny, M., & Grosch, W., (2000). Potent odorants of raw Arabica coffee. Their changes during roasting, *Journal of Agricultural and Food Chemistry, 48*, 868–872.

31. Czerny, M., Mayer, F., & Grosch, W., (1999). Sensory study on the character impact odorants of roasted Arabica coffee. *Journal of Agricultural and Food Chemistry, 47,* 695–699.

32. Daglia, M., Cuzzoni, M. T., & Dacarro, C., (1994). Antibacterial activity of coffee. *Journal of Agricultural and Food Chemistry, 42,* 2270–2272.

33. Daglia, M., Papetti, A., Gregotti, C., Berté, F., & Gazzani, G., (2000). *In vitro* antioxidant and ex vivo protective activities of green and roasted coffee. *Journal of Agriculture and Food Chemistry, 48,* 1449–1454.

34. Daniels, N., (2009). Variations in coffee processing and their impact on quality and consistency. M. Sc. (Forestry) Thesis, School of Forest Resources and Environmental Science, Michigan Technological University. Houghton, Michigan, pp. 140.

35. De Azevedo, A. B. A., Kieckbush, T. G., Tashima, A. K., Mohamed, R. S., Mazzafera, P., & Vieira de Melo, S. A. B., (2008). Extraction of green coffee oil using supercritical carbon dioxide. *Journal of Supercritical Fluids, 44,* 186–192.

36. Del Castillo, M. D., Gordon, M. H., & Ames, J. M., (2005). Peroxyl radical-scavenging activity of coffee brews. *European Food Research and Technology, 221,* 471–477.

37. Dutra, E. R., Oliveira, L. S., Franca, A. S., Ferraz, V. P., & Afonso, R. J. C. F., (2001). A preliminary study on the feasibility of using the composition of coffee roasting exhaust gas for the determination of the degree of roast. *Journal of Food Engineering, 47,* 241–246.

38. Etienne, H., (2005). Somatic embryogenesis protocol: coffee (*Coffea arabica* L. and *C. canephora* P.). In: Jain, S. M., Gupta, P. K. Eds., Protocol for Somatic Embryogenesis in Woody Plants. Volume 77 of the series Forestry Sciences, Springer Netherlands, pp. 167–179.

39. Farah, A., (2012). Coffee Constituents. In: Chu, Y. F. Ed., Coffee: Emerging Health Effects and Disease Prevention. 1st Ed., John Wiley & Sons, Inc. Published, Blackwell Publishing Ltd., pp. 22–54.

40. Farah, A., de Paulis, T., Trugo, L. C., & Martin, P. R., (2006). Chlorogenic acids and lactones in regular and water-decaffeinated *Arabica Coffee*. *Journal of Agricultural and Food Chemistry, 54,* 374–381.

41. Farah, A., & Donangelo, C. M., (2006). Phenolic compounds in coffee. *Brazilian Journal of Plant Physiology, 18,* 23–36.

42. Farah, A., Monteiro, M. C., Calado, V., Franca, A. S., & Trugo, L. C., (2006b). Correlation between cup quality and chemical attributes of Brazilian coffee. *Food Chemistry, 98,* 373–380.

43. Feria-Morales, A. M., (2002). Examining the case of green coffee to illustrate the limitations of grading systems/expert tasters in sensory evaluation for quality control. *Food Quality and Preference, 13,* 355–367.

44. Ferraz, M. B. M., Farah, A., Iamanaka, B. T., Perrone, D., Copetti, M. V., & Marques, V. X., et al., (2010). Kinetics of ochratoxin a destruction during coffee roasting. *Food Control, 21,* 872–877.

45. Flament, I., (2001). Coffee Flavor Chemistry. John Wiley & Sons, New York, NY, pp. 321.

46. Flament, I., Gautschi, F., Winter, M., Willhalm, B., & Stoll, M., (1968). Les composants furanniques de larome cafe: quelques aspects chimiques et spectroscopiques.

Proceedings of 3ʳᵈ International Scientific Colloquium on Green and Roasted Coffee Chemistry. ASIC, Paris, pp. 197–215.

47. Franca, A. S., Mendonça, J. C. F., & Oliveira, S. D., (2005). Composition of green and roasted coffees of different cup qualities. *LWT Food Science and Technology, 38*, 709–715.

48. Franca, A. S., & Oliveira, L. S., (2009). Coffee processing solid wastes: Current uses and future perspectives. In: Ashworth, G. S., Azevedo, P., Eds., Agricultural Wastes. Nova Publishers, New York, pp. 155–189.

49. Français, C., (1997). Coffee was the Discovery of coffee (Café a la découverte du café) Paris: Adexquation Publicite.

50. Franková, A., Drábek, O., Havlík, J., Száková, J., & Vanek, A., (2009). The effect of beverage preparation method on aluminium content in coffee infusions. *Journal of Inorganic Biochemistry, 103*, 1480–1485.

51. Fujioka, K., & Shibamoto, T., (2008). Chlorogenic acid and caffeine contents in various commercial brewed coffees. *Food Chemistry, 106*, 217–221.

52. Fukushima, Y., Ohie, T., & Yonekawa, Y., (2009). Coffee and green tea as a large source of antioxidant polyphenols in the japanese population, *Journal of Agricultural and Food Chemistry, 57*, 1253–1259.

53. George, S. E., Ramalakshmi, K., & Mohan Rao, L. J., (2008). A perception on health benefits of coffee. *Food Science & Nutrition, 48*(5), 464–486.

54. Ginz, M., Balzer, H. H., Bradbury, A. G. W., & Maier, H. G., (2000). Formation of aliphatic acids by carbohydrate degradation during roasting of coffee. *European Food Research and Technology, 211*, 404–410.

55. Gonzalez-Rios, O., Suarez-Quiroza, M. L., Boulanger, R., Barel, M., Guyot, B., & Guiraud, J. P., et al., (2007). Impact of "ecological" post-harvest processing on coffee aroma: I. Green coffee. *Journal of Food Composition and Analysis, 20*, 289–296.

56. Gonzalez-Rios, O., Suarez-Quiroza, M. L., Boulanger, R., Barel, M., Guyot, B., Guiraud, J. P., et al., (2007). Impact of "ecological" post-harvest processing on coffee aroma: II. Roasted coffee. *Journal of Food Composition and Analysis, 20* (3–4), 297–307.

57. Grembecka, M., Malinowska, E., & Szefer, P., (2007). Differentiation of market coffee and its infusions in view of their mineral composition. *Science of The Total Environment, 383*, 59–69.

58. Hernández, J. A., Heyd, B., & Trystram, G., (2008). On-line assessment of brightness and surface kinetics during coffee roasting. *Journal of Food Engineering, 87*, 314–322.

59. Higdon, J. V., & Frei, B., (2006). Coffee and health: A review of recent human research. *Critical Reviews in Food Science and Nutrition, 46*, 101–123.

60. Huxley, P. A., (1964). Some factors which can regulate germination and influence viability of coffee seeds. *Proceedings of International Seed Testing Association, 29*, 33–60.

61. Illy, A., & Viani, R., (2005). Espresso coffee: The science of quality. Academic Press, London, UK. Pp. 324.

62. Janıcek, G., & Pokorny, J., (1970). Changes of coffee lipids during the storage of coffee beans. *Zeitschrift fur Lebensmitteluntersuchung und Forschung A, 144*, 189–191.

63. Knopp, S., Bytof, G., & Selmar, D., (2006). Influence of processing on the content of sugars in green Arabica coffee beans. *European Food Research and Technology, 223*(2), 195–201.

64. Kolling-Speer, L., & Speer, K., (2005). The raw seed composition. In: Illy, A., Viani, R. Eds., Espresso Coffee, the Science of Quality. Elsevier Academic Press, Italy, pp. 148–178.

65. Kuit, M., Jansen, D. M., & Thiet, N. V., (2004). Manual for Arabica Cultivation of Coffee. Tan Lam Agricultural Product Joint Stock Company, Khe Sanh, Vietnam, pp. 212.

66. Kumazawa, K., & Masuda, H., (2003). Investigation of the change in the flavor of a coffee drink during heat processing. *Journal of Agricultural and Food Chemistry, 51*, 2674–2678.

67. Laderach, P., (2007). Management of intrinsic quality characteristics for high value specialty coffees of heterogeneous hillside landscapes. University of Bonn, Germany, pp. 110.

68. Leroy. T., Fabienne, R., Benoit, B., Pierre, C., Magali, D., & Christophe, M., et al., (2006). Genetics of coffee quality. *Brazilian Journal of Plant Physiology, 18*(1), 229–242.

69. Lima, D. R., (2003). Coffee manual of clinical pharmacology. Rio de Janeiro: Medsi Editora, pp. 321.

70. Mazzafera, P., Baumann, T. W., & Shimizu, M. M., (2009). Decaf and the steeplechase towards decaffito the coffee from caffeine-free Arabica plants. *Tropical Plant Biology, 2*, 63–76

71. Michael, D., (2016). Discovering coffee's unique health benefits, Life Extension Magazine, 2012. Accessed on 20 July URL: http://www.lifeextension.com/magazine/2012/1/Discovering-Coffees-Unique-Health-Benefits/Page-01.

72. Murekezi, A. K., (2003). Profitability analysis and strategic planning of coffee processing and marketing in Rwanda: a case study of a coffee growers' association. Department of Agricultural Economics, Michigan State University, East Lansing, Michigan, pp. 50.

73. Mussatto, S. I., Dragone, G., & Roberto, I. C., (2006). Brewer's spent grain: generation, characteristics and potential applications. *Journal of Cereal Science, 43*, 1–14.

74. Nabais, J. M. V., Nunes, P., Carrott, P. J. M., Carrott, M. R., García, A. M., & Díez, M. A. D., (2008). Production of activated carbons from coffee endocarp by CO_2 and steam activation. *Fuel Processing Technology, 89*, 262–268.

75. Navarini, L., Nobile, E., Pinto, F., Scheri, A., & Suggi-Liverani, F., (2009). Experimental investigation of steam pressure coffee extraction in a stove-top coffee maker. *Applied Thermal Engineering, 29*, 998–1004.

76. Navarini, L., & Rivetti, D., (2010). Water quality for Espresso coffee. *Food Chemistry, 122*, 424–428.

77. Nicoli, M. C., Anese, M., Manzocco, L., & Lerici, C. R., (1997). Antioxidant properties of coffee brews in relation to the roasting degree. *LWT – Food Science and Technology, 30*, 292–297.

78. Njoroje, J. M., (2004). Agronomic and Processing Factors Affecting Coffee Quality. *Outlook on Agriculture, 27*(3), 163–166.

79. Ogita, S., Uefugi, H., Yamaguchi, Y., Koizumi, N., & Sano, H., (2003). RNA interference: Producing decaffeinated coffee plants. *Nature, 423*, 823. doi:10.1038/423823a.

80. Ortiz, A., Veja, F. E., & Posada, F., (2004). Volatile composition of coffee berries at different stages of ripeness and their possible attraction to the coffee berry borer hypothenemus hampei (Coleoptera: Curculionidae). *Journal of Agricultural and Food Chemistry, 52*, 5914–5918.

81. Pandey, A., Soccol, C. R., Nigam, P., Brand, D., Mohan, R., & Roussos, S., (2000). Biotechnological potential of coffee pulp and coffee husk for bioprocesses. *Biochemical Engineering Journal, 6*, 153–162.

82. Parras, P., Martínez-Tomé, M., Jiménez, A. M., & Murcia, M. A., (2007). Antioxidant capacity of coffees of several origins brewed following three different procedures. *Food Chemistry, 102*, 582–592.

83. Pellegrini, N., Serafini, M., Colombi, B., Del Rio, D., Salvatore, S., & Bianchi, M., et al., (2003). Total antioxidant capacity of plant foods, beverages and oils consumed in Italy assessed by three different in vitro assays. *Journal of Nutrition, 133*, 2812–2819.

84. Peterson, A. S., (2007). Health benefits of coffee. *The Journal of Lancaster General Hospital, 2*(4), 146–147.

85. Petracco, M., (2001). Beverage preparation: Brewing trends for the new millennium. In: Clarke, R., Vitzthum, O. Eds., Coffee: Recent Developments. Oxford: Blackwell Science, pp. 230.

86. Pokorny, J., Con, N. H., Smidrkalova, E., & Janıcek, G., (1975). Non-enzymic browning. XII. maillard reactions in green coffee beans on storage. *Zeitschrift fur Lebensmitteluntersuchung und Forschung, 158*, 87–92.

87. Ramalakshmi, K., Rao, J. M., Takano-Ishikawa, Y., & Goto, M., (2009). Bioactivities of low-grade green coffee and spent coffee in different in vitro model systems. *Food Chemistry, 115*, 79–85.

88. Ratnayake, W. M. N., Hollywood, R., O'Grady, E., & Stavric, B., (1993). Lipid content and composition of coffee brews prepared by different methods. *Food and Chemical Toxicology, 31*, 263–269.

89. Reh, C. T., Gerber, A., Prodoillet, J., & Vuataz, G., (2006). Water content determination in green coffee method comparison to study specificity and accuracy. *Food Chemistry, 96*, 423–430.

90. Rick, H., & Graham, F., (2004). Crop post-harvest: Science and technology, Durables case studies in the handling and storage of durable commodities, volume 2. Blackwell Science Ltd, pp. 251.

91. Rojas, J., (2004). Green coffee storage. In: Wintgens, J. N. Ed., Coffee: Growing, Processing, Sustainable production. Weinheim: Wiley-VCH, pp. 733–750.

92. Sacchetti, G., Di Mattia, C., Pittia, P., & Mastrocola, D., (2009). Effect of roasting degree, equivalent thermal effect and coffee type on the radical scavenging activity of coffee brews and their phenolic fraction. *Journal of Food Engineering, 90*, 74–80.

93. Saura-Calixto, F., & Goni, I., (2006). Antioxidant capacity of the Spanish Mediterranean diet. *Food Chemistry, 94*, 442–447.

94. Scheidig, C., & Schieberle, P., (2006). Einfluss der lagerung von rohkaffee auf das aroma von rohkaffee, Rostkaffee und kaffeegetrank, *Lebensmittelchemie, 60*, 55–56.

95. Selmar, D., Bytof, G., Knopp, S. E., Bradbury, A., Wilkens, J., & Becker, R., (2004). Biochemical insights into coffee processing: Quality and nature of green coffees are interconnected with an active seed metabolism. 20ème Colloque Scientifique International sur le Café; 11–15 Bangalore, India and Paris: Association Scientifique Internationale du Café (ASIC), 2005, 262.

96. Selmar, D., Bytof, G., Knopp, S. E., & Breitenstein, B., (2006). Germination of coffee seeds and its significance for coffee quality. *Plant Biology, 8,* 260–264.

97. Shlonsky, A. K., Klatsky, A., & Armstrong, A., (2003). Traits of persons who drink decaffeinated coffee. *Annals of Epidemiology, 13,* 273–279.

98. Silva, E. A., Mazzafera, P., Brunini, O., Sakai, E., Arruda, F. B., & Mattoso, L. H. C., et al., (2005). The influence of water management and environmental conditions on the chemical composition and beverage quality of coffee beans. *Brazilian Journal of Plant Physiology, 17*(2), 229–238.

99. Silva, M. A., Nebra, S. A., Machado Silva, M. J., & Sanchez, C. G., (1998). The use of biomass residues in the Brazilian soluble coffee industry. *Biomass and Bioenergy, 14,* 457–467.

100. Silvarola, M. B., Mazzafera, P., & Fazuoli, L. C., (2004). A naturally decaffeinated arabica coffee. Nature, pp. 249–826.

101. Sivetz, M., & Desrosier, N. W., (1979). Hulling, classifying, storing, and grading green coffee beans. In: Sivetz, M., Desrosier, N. W. Eds., Coffee technology. The avi Publishing company, Westport, CT, pp. 117–169.

102. Smith, A. W., (1987). Introduction. In: Clarke, R. J., Macrae, R. Eds., Coffee, volume 1: Chemistry. Elsevier: Applied Science Publishers, London, pp. 1–44.

103. Speer, K., & Kolling-Speer, I., (2006). The lipid fraction of the coffee bean, *Brazilian Journal of Plant Physiology, 18,* 201–216.

104. Susila, W. R., (2005). Enhancement of coffee quality through prevention of mould formation: targeted study of the arabica coffee production chain in north sumatra (The Mandheling Coffee), National consultant report, food and agriculture organization, United Nations, pp. 97.

105. Svilaas, A., Sakhi, A. K., Andersen, L. F., Svilaas, T., Strom, E. C., & Jacobs, D. R., et al., (2004). Intakes of antioxidants in coffee, Wine and vegetables are correlated with plasma carotenoids in humans. *Journal of Nutrition, 134,* 562–567.

106. Taniwaki, M. H., Pitt, J. I., Teixeira, A. A., & Iamanaka, B. T., (2003). The source of ochratoxin a in Brazilian coffee and its formation in relation to processing methods. *International Journal of Food, Microbiology, 82,* 173–179.

107. Toci, A. T., & Farah, A., (2008). Volatile compounds as potential defective coffee seeds' markers. *Food Chemistry, 108*(3), 1133–1141.

108. Toci, A. T., Farah, A., & Trugo, L. C., (2006). Efeito do processo de descafeinac,ao com diclorometano sobre a composic,ao quimica dos cafes Arabica e Robusta antes e apos a torracao, *Quimica Nova, 29*(5), 965–971.

109. Tokimoto, T., Kawasaki, N., Nakamura, T., Akutagawa, J., & Tanada, S., (2005). Removal of lead ions in drinking water by coffee grounds as vegetable biomass. *Journal of Colloid and Interface Science, 281,* 56–61.

110. Trugo, L. C., & Macrae, R., (1984). A study of the effect of roasting on the chlorogenic acid composition of coffee using HPLC. *Food Chemistry, 15,* 219–227.

111. Trugo, L., (2003). Coffee. In: Caballero, B., Trugo, L., Finglas, P. (Eds.), Encyclopaedia of Food Sciences and Nutrition. 2nd Ed., Academic Press, London, pp. 765.

112. Vaast, P., Betrand, B., Perriot, J. J., Guyot, B., & Genard, M., (2006). Fruit thinning and shade improve bean characteristics and beverage quality of coffee (*Coffea arabica L.*) under optimal conditions. *Journal of the Science of Food and Agriculture, 86*, 197–204.

113. Variyar, P. S., Ahmad, R., Bhat, R., Niyas, Z., & Sharma, A., (2003). Flavoring components of raw monsooned Arabica coffee and their changes during radiation processing. *Journal of Agricultural and Food Chemistry, 51*, 7945–7950.

114. Varnam, A., & Sutherland, J. M., (1994). Beverages: Technology, Chemistry and Microbiology. Springer Science & Business Media, pp. 464.

115. Veedhi, A., (2008). Coffee cultivation guide for south - west monsoon area growers in India (coffee kaipidi). Central Coffee Research Institute, Karnataka, India, pp. 231.

116. Velasco, J. R., & Guitierrez, J., (1974). Germination and its inhibition in coffee. *Philippine Journal of Science, 103*, 1–11.

117. Viani, R., (2001). Global perspectives in coffee quality improvement. In: Chapman, K., Subhadrabandhu, S. Eds., The First Asian Regional Round-Table on Sustainable, Organic and Specialty Coffee Production, Processing and Marketing. 26-28th February, Chiang Mai, Thailand, pp. 110–115.

118. Willson, K. C., (1999). Coffee, Cocoa, and Tea. CABI Publishing, New York, pp. 300.

119. Wurziger, J., Drews, R., & Suche, B., (1982). Uber rust from deposited green coffee (*Uber Rostkaffees aus abgelagerten Rohkaffees*). *Kaffee und Tee Markt, 32*, 3–5.

CHAPTER 5

BIOCHEMICAL COMPOSITION, PROCESSING TECHNOLOGY AND HEALTH BENEFITS OF GREEN TEA: A REVIEW

VISHAL SINGH, DEEPAK KUMAR VERMA, and
DIPENDRA KUMAR MAHATO

CONTENTS

5.1   INTRODUCTION

Green tea obtained from *Camellia sinensis* belongs to the genus *Camellia*, a member of the family *Theaceae* originated in China [97] but it has become associated with many cultures throughout Asia and is spread throughout different parts of world. It is one of the most popular beverages, consumed by over two-thirds of the world's population, next only to water. Green tea is grown in over 30 countries, exclusively in the subtropical and tropical zones [47]. It is estimated that about 2.5 million tons of tea leaves are produced each year throughout the world, in which green tea accounts for approximately 20% of total produced and consumed in the world mainly in Asian countries including India, some parts of North Africa, the United States, Europe, China and Japan [26, 80, 118] as one of the most ancient and popular therapeutic beverages.

Many varieties of green tea have been cultivated in the countries where it is grown. In India, it is cultivated in different parts of Assam, Darjeeling, Travancore, the Nilgiris, Malabar, Bengal, Dehra Dun and Kumaon [72]. Among all types of tea (categorized into three types, depending on the level of fermentation: green (unfermented), oolong (partially fermented) and black (fermented) tea), green tea has more catechins (anti-oxidants) because green tea leaves are steamed not fermented. Hence it preserves more polyphenols mainly catechins, which contain nearly 30% of dry weight of green tea leaves. Study on green tea revealed that iron content present in consumable food reduced the beneficial effect of the anti-oxidants in green tea [92].

Green tea is a one of the most popular beverage in entire globe followed by coffee, beer and wine [26, 44, 71, 110, 134, 135]. As per fermentation level, tea has been classified as without fermentation as fresh (green tea), slightly fermented (oolong tea), completely fermented (black tea) as well as tea made with dried leaves and newly emerged buds after steaming (due to steaming, polyphenol oxidation has been deactivated) [104]. Bud and newly emerged leaves of green tea were treated as richest source of epigallocatechin gallate [44, 66]. Besides beneficial for human health, green tea is effective in reducing the chances of tumors and carcinogenesis [33]. It is also noted that anti-oxidants of green tea have beneficial impact in the case of red blood cell haemolysis [155].

This chapter reviews status, importance, biochemical composition, processing technologies and health benefits of green tea.

5.2 GREEN TEA: BOTANICAL DESCRIPTION AND DIVERSITY

Green tea is an evergreen shrub or tree (Figure 5.1) and can grow up to a height of 30 feet (10 meters), but is usually pruned to 2–5 feet for cultivation. It has a strong taproot. The flowers are yellow-white, 2.5–4 cm in diameter, with 7 to 8 petals. The leaves are dark green, alternate and oval (4–15 cm long and 2–5 cm broad) with serrated edges, and the blossoms are white, fragrant, and appear in clusters or singly. The young, light green leaves are preferably harvested for tea production; they have short white hairs on the underside. However, the older leaves are deeper green. Differences in tea qualities are due to different chemical compositions at leaf ages. Furthermore, species of green tea add more varieties with geographical, ecological and chemical features, and have been grown in variable growing conditions: those grown in under direct sun and another one grown under the shade. These varieties can differ substantially due to variable growing conditions, horticulture, production processing, and harvesting time. Scientific classification is shown below:

Kingdom	Plantae
Order	Ericales
Family	Theaceae
Genus	Camellia
Species	C. sinensis
Binomial Name	*Camellia sinensis* (L.) Kuntze

A)   B)

FIGURE 5.1   Green tea (*Camellia sinensis*): (A) Leaves, and (B) Fruits.

A wide variation is found in the diversity of green tea, which varies from country to country and is also known by specific name in the respective country. Primarily, two basic varieties of tea have been recognized as Chinese variety (*C. sinensis* Sinensis) and Assamese variety (*C. sinensis* Assamica). From ancient time, tea was used as a beverage in China and it was later introduced in Europe in 16th century. In 19th century (1823 A.D.), cultivation of tea plant started first time in India (region of Assam) [94]. Tropical and subtropical climate is suitable for tea cultivation with varying range of other phenomena like temperature range (13°C–29°C), altitude (2460 m above sea level), and soil characteristics (containing high amount of iron and manganese, pH range of 3.3–6.0: acidic) [94]. Green tea is one of the most popular drinks and millions of cups are being consumed every day in the entire world [91].

The plant was originally discovered and grown in South-east Asia about 1000 years ago and according to the Chinese mythology, the Emperor Shen Nung discovered tea for first time in 2737 B.C. [51]. In 1753 A.D., Carl Linnaeus in his "Species Plantarum" (1st ed.) nominated *Camellia* cultivated in Japan as *Camellia japonica* and the tea plants cultivated in China as *Thea sinensis*, respectively [145]. Abundant biodiversity has been found in the wild population of *C. sinensis*. It is because of variation in chromosome structure that reveals heterozygosity and polymorphism. The variability of morphology is expressed mainly by the differentiation on the basis of form, size, texture, color and vesture of leaves, sepals and petals [145].

5.3 BIOCHEMICAL COMPOSITION OF GREEN TEA

Green tea indicates the tea that is prepared with the fresh green leaves of tea plant [129]. The biochemical composition of green tea includes different constituents such as (Table 5.1): protein and amino acids; fiber; carbohydrates; lipids; vitamins and anti-oxidants; pigments; volatile compounds; minerals; phenolic compounds (like polyphenols and flavonoids) [105]. The presence of certain minerals and vitamins (like vitamin-C and vitamin-E) increase the anti-oxidant potential of green tea. Alkaloids, saponins, tannins, catechin and polyphenols were screened by Mbata et al. [89] and revealed the presence of phytochemicals in green tea. Chen and

TABLE 5.1 Biochemical Composition of Green Tea

Constituents	Remarks	Reference
Amino acids	Amino acids (1–4% dry weight) such as thiamine or 5-N-ethylglutamine, glutamic acid, tryptophan, glycine, serine, aspartic acid, tyrosine, valine, leucine, threonine, arginine, and lysine.	[20, 44]
Carbohydrates	Carbohydrates (5–7% of dry weight) such as: cellulose, pectin, glucose, fructose and sucrose.	[20]
Lipids	Lipid components: linoleic and linolenic acids and sterols such as stigma sterol.	[20]
Minerals	Minerals elements around 5% on the dry weight basis (such as Ca, Mg, Cr, Mn, Fe, Cu, Zn, Mo, Se, Na, P, Co, Sr, Ni, K, F and Al) can be found in green tea.	[4, 15, 20]
Pigments	Pigments (such as chlorophyll and carotenoids) are present.	[20]
Polyphenols	Epicatechin (EC), Epigallocatechin (EGC), Epicatechin-3-Gallate (ECG) and Epigallocatechin gallate (EGCG) are identified in green tea.	[112, 115]

Zhou [23] found polyphenols, caffeine, amino acids, and minerals. The main constituents vary from cultivar to cultivar in tea due to differences in variety, origin, and growing conditions [73]. All chemical constituents increase gradually among: shoots with one leaf and a bud, second leaf, third leaf, and fourth leaf [136]. Agricultural practices (viz. fertilizers, minerals, soil, water, etc.), species of the plant, growing season, age of the leaf, climatic conditions are factors that affect the biochemical composition of green tea.

Polyphenols are the secondary metabolites of plants and have a great potential as an alternative source of treatment of chronic diseases. Over 4000 different flavonoids have been described [39], and they are categorized into flavonols, flavones, catechins, flavanones, anthocyanidins and isoflavonoids. Phenolic compounds are phytochemical screening of green tea and may account for up to 30% of the dry weight [90], which reveals the presence of flavanols (such as kaempferol, myricetin and quercetin are commonly known as catechins); flavandiols; flavonoids; and phenolic acids such as gallic acid, chlorogenic acid, caffeic acid [44]. Catechins are colorless, water-soluble compounds that contribute to the bitterness and astringency of green tea and are found in greater and most abundant amounts [134]; and it is approximately 90 mg of a typical green tea [7] greater than in black tea (4.3%). It is noticed that higher amounts of catechins are present in green tea than black and oolong tea, because of differences in the processing of tea leaves after harvest. Sano et al. [112, 115] used high performance liquid chromatography (HPLC) to identify four types of catechins in green tea (Figure 5.2): (i) epicatechin (EC); (ii) epigallocatechin (EGC); (iii) epicatechin-3-gallate (ECG); and (iv) Epigallocatechin gallate (EGCG). Naghma and Hasan [96] reported that polyphenol and its content was related to its catechin, particularly EGCG content of about 60–70% in the total catechins [65] in green tea showing attributes of health-promoting effects [20].

Catechins (responsible for bitterness and astringency in tea) contain high anti-oxidant activity with anti-oxidative, anti-tumor and anti-inflammatory effects, thus beneficial to human health [13, 47, 88]. They also

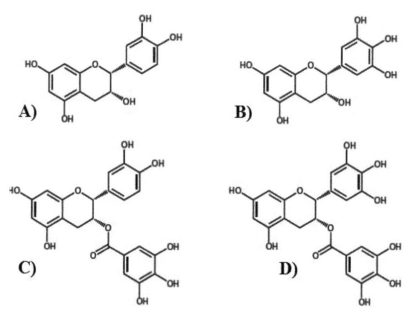

FIGURE 5.2   Chemical structures of four major catechins presents in green tea; (A) Epicatechin (EC); (B) Epigallocatechin (EGC); (C) Epicatechin-3-Gallate (ECG); and (D) Epigallocatechin gallate (EGCG).

exhibited anti-obesity effect in animals [29, 93, 101, 127, 133] and humans [50, 130]. It mainly comprises of (–)-epigallocatechin gallate (EGCG), (–)-epigallocatechin (EGC), (–)-epicatechin gallate (ECG), (–)-epicatechin (EC), and their geometric isomers (–)-gallocatechin gallate (GCG), (–)-gallocatechin (GC), (–)-catechin gallate (CG), and (–)-catechin(C), among which, EGCG is regarded as the most important catechin due to high content in tea leaf and excellent bioactivity [38, 64, 111, 133]. The scavenging abilities of EGCG and GCG are higher than those of non-gallated catechins EGC, GC, EC and C due to the presence of a gallate moiety at C-3 position [46]. Han et al. [48] and Huvaere et al. [58] have reported that catechins affect the composition of curds and cheese due to the interactions between milk components and catechins. These catechins interact non-covalently with proteins [154] or with the milkfat globules [82, 152]. These challenges can be overcome by nano-encapsulation to protect catechins from interacting with milk components [108]. It is reported that tea catechins inhibit the activity of α-amylase and α-glucosidase, which are essential for starch digestion in humans [74, 86, 153]. Among the 4 types of digestive enzymes (*viz.*, α-amylase, pepsin, trypsin and lipase), α-amylase activity is inhibited the most [52]. Catechins with galloyl group is a natural tyrosine kinase inhibitor that a signalling kinases, ERK1/2, proteinkinase B (Akt), PI3K and p38 mitogen activated protein kinase [123]. Phenolic hydroxyl group of catechins can inhibit lipid peroxidation and fat hydrolysis while galloyl group contributes to the prostacyclin production, reduction of vascular cell adhesion molecule-1 expression [5].

5.4 CONSUMPTION OF GREEN TEA

Tea has been associated with our daily life for a long historical period [145]. The worldwide consumption of tea per capita is 120 ml brewed tea per day [2]. Approximately 80% of the tea is manufactured and consumed globally as black tea in the world, compared to 20–22% of green tea and less that 2% of oolong tea. Black tea is consumed principally in Europe, North America and North Africa while green tea is widely drunk in China, Japan, Korea, and Morocco. Oolong tea is popular in China and Taiwan [15]. Green tea contains polyphenols, which include epigallocatechin gallate (EGCG), epicatechin gallate ($C_{22}H_{18}O_{10}$), epicatechins ($C_{15}H_{14}O_{6}$); and flavanols ($C_{15}H_{14}O_2$) [69], which provide numerous health benefits like prevention from cancer, leukemia, liver disease, diabetes [104].

Studies revealed that regular intake of green tea reduces the risk of oesophageal cancer (in Chinese population), lung cancer (in women), oral cancer (in Asian people), liver cancer (in Chinese and Japanese people) [11, 40, 57, 69, 138, 139, 158]. However, people getting medicated with bortezomib drugs used for chemotherapy and proteasome inhibitors based on boronic acid should avoid intake of green tea, since it can interfere with these drugs [61]. It also helps in curing dental caries and infections resulting from nutrition and invasion due to bacteria. It reduces damaging of cell membranes, helps in preventing cell damage in oral cavity due to virus and enhances enzymatic activity of the body [99]. Ingestion of bioactive components available in green tea may cause decrease in loss of bones resulting to osteoporosis [117].

Caffeine is also a bioactive chemical compound (Figure 5.3) that is found in tea leaves and well known for central nervous system (CNS) stimulant of the methylxanthine class [100]. Due to its stimulating action and anti-cancerous role, it is also popular in medicine and pharmaceutical industries. This is the main reason that people (generally, aged person and pregnant women) use decaffeinated tea if they suffer with insomnia or sleep disruption (refer as sleeping difficulty) [1, 14, 60, 119]. Furthermore, researchers of this area recommend exploring the consumption of green tea because it is also associated with lower risk of chronic disease *viz.* Parkinson's disease, cardiovascular diseases (CVD) and cancer, etc. (Table 5.2) [18, 35, 63, 106]. Consumption of green tea discourages the effect of different diseases like cancer [70]; high and low blood pressures as well as

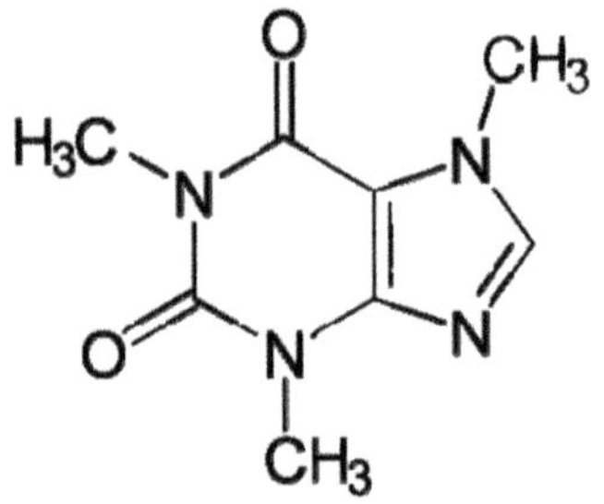

FIGURE 5.3   Chemical structure of bioactive compound caffeine found in green tea.

TABLE 5.2 Medicinal and Pharmacological Effects of Green Tea on Human Health

Effects	Remarks	References
Anti-aging activity	• Cancer, Parkinson's disease, Alzheimer's disease, CVDs and diabetes are diseases associated with several age and have their etiologies linked to changes in oxidant/anti-oxidant balances and free radical damage.	[3]
	• Although there is no epidemiological evidence in human studies of the benefit of green tea for Alzheimer's disease, but several studies in animal suggested that EGCG from green tea may affect several potential targets associated with Alzheimer's disease progression.	
Anti-Alzheimer activity	• Although there is no epidemiological evidence in human studies of the benefit of green tea for Alzheimer's disease, but several studies in animal suggested that EGCG from green tea may affect several potential targets associated with Alzheimer's disease progression.	[3]
Anti-Parkinson activity	• Green tea significantly prevent to the Parkinson's disease. In Asian populations, the prevalence of Parkinson's disease and green tea consumption show lower (5-10 folds) incidences of the disease.	[6, 8]
Anti-stroke activity	• Increased consumption of green tea reduces risk from stroke. Chances for suffering stroke becomes less (21%) if consumes three or more cups of green tea.	[9]
Cardiovascular diseases	• The protective effect of green tea in CVDs is also thought to stem from its anti-oxidant activity. Indeed, oral intakes of green tea extract by human volunteers have an effect that may lower the risk of atherogenesis. However, the potential benefits of tea consumption are worthy of confirmation by more experimental researches	.[10]
Anti-cancer activity	• Green tea has a reputed role in cancer prevention as tea catechins have been shown to inhibit tumour cell proliferation as well as promote the destruction of leukaemia cells.	[12]
Anti-diabetic activity	• Green tea has an anti-diabetic effect. It lowered glucose levels in the bloodstreams of diabetic without affecting insulin levels. Green tea extract was also found to prevent development of insulin resistance, hyperglycaemia and other metabolic defects.	[16, 17]

TABLE 5.2   (Continued)

Effects	Remarks	References
Anti-caries Activity	• Green tea consumption is one of the most practical cancer preventives which prevent many types of cancer, including lung, colon, esophagus, mouth, stomach, small intestine, kidney, pancreas, and mammary glands.	[19, 21, 22]
	• Green tea extract is effective in preventing dental caries. Tea leaves are rich in fluoride, which is known to enhance dental health and prevents dental caries. However, the possible dental health benefits of tea are not limited to fluoride, but involve other tea components (like polyphenols), have preventative effects on dental caries.	
Obesity and Weight Loss	• Obesity has increased at an alarming rate in recent years and is now a worldwide health problem. Consumption of green tea protects against obesity-related disorders such as artherosclerosis, diabetes and hypertension.	[27]

cardiovascular risk [34], besides its good impact on immune function and liver injury because of presence of amino acid in green tea leaves that is L-theanine that produces "umami", a type of flavor [156].

5.5   PROCESSING TECHNOLOGIES OF GREEN TEA

Tea is one of the three major non-alcoholic beverages in the world [147]. The processing of green tea is done with the help of different equipment (Figure 5.4) through various steps (Figure 5.5).

Freshly harvested young and unfermented leaves are immediately steamed to prevent fermentation, yielding a dry, stable product for producing green tea. A slight amount of indoor withering or exposure to diffused sunlight is allowed to the tea. This steaming process destroys the enzymes responsible for breaking down the color pigments in the leaves and allows the tea to maintain its green color during the subsequent rolling and drying processes. During the processing, green teas are not fermented and thus retain the original color of the tea leaves because of the oxidizing enzymes [26].

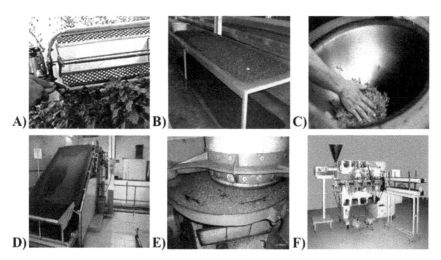

FIGURE 5.4 Equipments used in different unit operations of green tea processing: (A) Plucking machine, (B) Withering process, (C) Pan frying process, (D) Drying process, (E) Rolling process, and (F) Packaging machine.

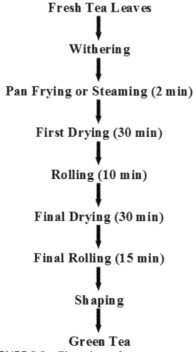

FIGURE 5.5 Flow chart of green tea processing.

The main enzymes in tea leaves are: polyphenol oxidase (PPO), catalase (CAT), peroxidase (PO), and ascorbic acid oxidase (AAO) [147], which are also responsible for various biochemical metabolic pathways. These processes preserve natural polyphenols with respect to the health-promoting properties. As green tea is fermented to oolong and then to black tea, polyphenol compounds (catechins) in green tea are dimerized to form a variety of theaflavins, so that these teas may have different biological activities [20]. Generally soon after harvest, inactivated enzymes cannot break down the tea tissue chlorophyll, which contributes to the green color.

Rolling is the next unit operation in which leaves are twisted, turned and broken with the roller machines and enzymes are released from the leaf as the leaf breaks which causes natural process of oxidation. Further, leaves are kept into the large tray and are allowed to oxidize by exposing them to air which generates heat, and slowly changes in color from green to red and then to brown due to oxidation process. After oxidation, leaves are exposed to hot air with help of air blowers for removing the remaining moisture of the leaf and finally dried up to recommended level for packing. Green tea is processed in different ways during manufacturing. For understanding the above technology, stepwise processing has been discussed here.

5.5.1 PLUCKING

Usually green tea leaves are plucked (generally when the green tea plants contain two or three leaves with bud), by experienced or skilled workers (who can be pluck around 40 kg of tea leaf per day); and are stored in a hanging basket or linen over their shoulder. Collected leaves should be weighed after sorting out the defective one and then good quality leaves are spread in trays or on the ground for 1–3 hours to remove undesired odor and reduce the moisture up to optimum level. Mature and uniform leaves with ideal practices of plucking contribute to quality of final and processed tea. Immature shoots and coarse leaves are unacceptable because standard flush depends upon leaves standard: Fine plucking, medium plucking or coarse plucking depending on >75%, 60–75%

or <60% of standard flush [120, 132]. Manual plucking of tea leaves has been labor intensive, time taking and less efficient, which draw the attention of researchers to develop an efficient, cheap and quick mechanized alternate to harvest the tea leaves even [113].

5.5.2 WITHERING

Withering is a unit operation in which drying of leaves is performed with help of natural or forced circulated air to reduce the moisture content and humidity. This process makes tea leaves soften, pliable and permeable, etc. [114], besides enhancement of aroma and flavor (Table 5.3) [94]. Withering of tea leaves is carried for ease of further unit operations like shaping and rolling. Development of aroma and flavor is directly proportional to

TABLE 5.3 Major Physical and Biochemical Changes During Each Processing Step

Physical and biochemical changes	Processing steps					
	Spreading	Withering	Rotating	Fixing	Rolling	Drying
Aroma transformation and release	X	**	X	**	x	**
Breakdown of chlorophyll	X	x	**	x	x	x
Disruption of cell walls	X	x	**	x	**	x
Emission of grassy odor	**	**	**	**	x	x
Fermentation	X	x	**	**	x	x
Hydrolysis	X	**	**	**	x	x
Moisture removal	**	**	**	**	x	**
Oxidation of catechins	X	**	**	**	x	**
Shape	X	x	**	x	**	x

LEGEND: x = No change; ** = There was a change

duration of withering due to formation of different volatile compounds during this unit operation. This process will take 12–24 hours depending on the nature of drying media (e.g., if the leaves are dried in the open air at an ambient temperature, more time will needed).

5.5.2.1 Physical Changes in Green Leaf During Withering

Reduction in moisture level begins just after plucking of leaves and it is continuously reduced up to desired level till completion of withering operation [77]. Permeability of cell membrane also increases with increased degree of withering.

5.5.2.2 Biochemical Changes in Green Leaf During Withering

Many biochemical changes have been observed during withering [31]. Tomlins and Mashingaidze [131] have noticed many changes in chemical compounds: proteins converted into amino acids, reduction of lipids, fatty acids, carotenoid and chlorophylls besides increment in caffeine content. Color, aroma, flavor, organic acid and volatile compounds are also altered during withering.

5.5.3 FIXING (PAN FRYING OR STEAMING)

Steaming or fixing is a unit operation and is performed for de-catalyzing the fermentation that causes changes in enzymes. However, it preserves the original color, texture as well as anti-oxidant compound of leaves (Table 5.3) [149]. In this process, leaves are poured in rotating drum and then are exposed to steam for inhibiting the oxidation [137]. During steaming, temperature of steam should be controlled and maintained between 100–180°C because excess (leaves will damage and undesired color, etc. will occurred) and less steaming (less impact on fermentation) can affect the quality of tea leaves. Moisture is removed from tea leaves during steaming and contains a significant amount of compounds like phenols, catachins and flavanols, etc. [7]. After steaming, tea leaves are formed into needle shape with dark green in color. Steamed green tea contains highest vitamin-C compared to black and oolong tea [97].

5.5.4 FIRST DRYING

First drying takes place just after completion of steaming for reducing the moisture content up to desired level (nearly 50% of moisture is removed during this unit operation). Generally leaves are dried with help of drum dryers (common drums constructed with wooden or metal alloys). During drying, the temperature of drying media (air) and drying time should be maintained around 55°C and half an hour, respectively.

Adequate drying retains the quality like aroma and taste of final tea [146]. Different methods of drying can be implemented like hot-air drying, vacuum drying, microwave drying and tray drying, etc. Lou [87] stated that microwave irradiation is applied for removal of moisture by heating the external and internal surface simultaneously [87]. Comparative study of different drying methods reveals that the order of rehydration ratio as: microwave vacuum drying > microwave drying > vacuum drying > hot-air drying. Gentle heating with the help of microwave vacuum drying resulted in best quality tea and also retained nutrient losses. On other hand, hot air drying destructed the cells of leaves due to direct application of more temperature and resulted in significant losses of vitamins, proteins and amino acids, etc. [85].

5.5.5 ROLLING

In this unit operation, leaves should be cleaned properly and then rolling machine is passed by at controlled pressure to press the leaves as these pass through rotor vane machine for fine crushing. After that, crushed leaves are transferred to curl-turn machine for obtaining finer particles. Finally, crushed tea leaves are passed through roller breaker to retard the fermentation. Adequate and proper rolling requires uniformity of the crushed particle, because excess rolling results in unbalanced mixing of chemical and enzyme (Table 5.3) [97].

5.5.6 FINAL DRYING, ROLLING, AND POLISHING

Final drying takes place just before the final rolling with help of a hot air for removal of moisture to desirable level at which rolling will be take place properly. For final rolling, dried tea leaves should be placed between heated plates. Appearance of tea leaves is improved due to final rolling.

5.6 FACTORS AFFECTING QUALITY OF GREEN TEA

Many factors affect the quality of green tea such as: type of processing technology, tea cultivar, shape of tea leaves, fertilization, season, location, climate, soil type and plucking standards, etc. Among these factors, processing factors have the greatest influence on the quality of green tea [30]. Leaf spreading for a short period before fixing is an important step for processing of high quality green tea. Spreading out of green leaves on bamboo trays or on the ground can promote the hydrolysis of non-water-soluble carbohydrates and pectins, formation and accumulation of non-gallated catechins, release of grass-like odor, and loss of some moisture in fresh leaves for better fixation. The spreading height, turning numbers, and duration for a particular type of green tea differ according to green leaf and weather conditions. Generally speaking, 70% of moisture content after spreading is suitable.

Deactivating the PPO is the main purpose of fixing. In China, a commonly accepted principle is "tender leaves need heavier fixing while mature leaves need lighter fixing; the fixing temperature should be higher at the beginning and then lower; and combination of promoting the moisture removal and inhibiting moisture removal". As tender leaves usually have higher enzyme activities, only higher temperature and long fixing time can thoroughly deactivate the enzymes. Heavier fixing of tender leaves will promote the hydrolysis of proteins. For example, the amino acid contents were increased with the longer fixing time [24]. However, over-fixing would scorch the leaf and results in a smoky taste and higher ratio of broken leaves. Lower temperature or shorter fixing time often produces red leaves due to the oxidation of polyphenols, which will lower the quality of green tea [26]. Otherwise, relatively lower temperature is needed for fixing mature leaves as they have lower water content and enzyme activities in comparison with tender leaves.

Rolling is also important for green tea quality. The degree of pressure and rolling time are key technical parameters. Longer rolling and heavier pressure imposed on fixed leaves will produce yellowish leaves and more broken leaves. Furthermore, hydrolysis of chlorophyll and auto-oxidation of polyphenols cause poor taste in green tea as more juices are squeezed out. Tea cultivars have also profound influence on green tea quality. Early-sprouting and high ratio of amino acid to polyphenols cultivars such as cv, Longjing 43, cv, Fuding Dabaicha (China cultivar) and cv. Yabukida (Japan cultivar), etc. usually produce high quality green tea as compared to other cultivars under the same cultivation and processing conditions.

Proper shading of the plant can also produce high quality green tea such as Gyokura. The tea plant is cultivated under the shade for about 2 weeks (60–90% darkness) before plucking the flush. Higher rates of nitrogen application has been shown to produce high quality green tea as amino acid levels in the leaves are increased compared to none or lower rates of nitrogen application. Fine plucking standard is undoubtedly favorable to green tea quality.

5.7 HEALTH BENEFITS OF GREEN TEA

In recent years, the benefits of consuming green tea and green tea constituents on our health are under investigation for prevention of cancer and CVDs; heart and liver disease; anti-inflammatory; anti-arthritic; anti-bacterial; anti-angiogenic; anti-oxidative; anti-viral; neuroprotective; and cholesterol-lowering effects by many researchers (Table 5.4) [20, 36, 49, 68, 90, 103, 107, 116, 124, 125, 140, 141] and also improvement of coronary flow velocity reserve [26].

5.7.1 DOSAGE OF GREEN TEA

Green tea is an important and daily worldwide consumed beverage. It is used from ancient time and got popularized progressively due to its flavor, aroma and medicinal characteristics. Tea leaves contain several chemical compounds, which have pharmaceutical importance. Proportion of different chemical compounds varies due to climatic conditions, variety of tea,

TABLE 5.4   Effects of Green Tea Consumption on Type of Cancer

Cancer Type	Authors/ researchers authors	Name of reports/ papers	Inference/remarks/ effect on health	Refer-ence
Breast cancer	Zhang et al.	Green tea and the prevention of breast cancer: A case control study in southeast China	The onset of breast cancer reduced among women, drinking GT (1–249 g of dried GT leaves yearly), regularly (OR=0.87; 95 % CI=0.73-1.04)	[157]
	Wu et al.	Green tea and risk of breast cancer in Asian Americans	There was a significant reduction in breast cancer among women who drank GT than did not. ORs adjusted for age and potential confiding factors. The risk of breast cancer further decreased with increasing dose of GT (OR=1.00, 0.71- 95 % CI=0.51-0.99) and (OR=0.53; 95% CI=0.35-0.78), respectively in association with number of GT cups/day (none, 0 85.7 mL, and > 85.7).	[142, 143]
Esophageal cancer	Gao et al.	Reduced risk of esophageal cancer associated with green tea consump-tion.	High consumption of GT reduced the risk of cancer in men and women who did not smoking cigarette/drink alcohol, High use: (OR= 0.83 95 % CI= 0.59-1.16 for men) and (OR= 0.29; 95 % CI=0.13-0.65 for women)	[43]
Lung cancer	Zhong et al.	A population-based case-control study of lung cancer and green tea consump-tion among women living in Shanghai, China	Regular consumption of GT reduce risk of lung cancer in non-smoker women (OR= 0.65; 95 % CI=0.45–0.93); and risk decreased with increasing doses of tea.	[159]

TABLE 5.4 (Continued)

Cancer Type	Authors/ researchers authors	Name of reports/ papers	Inference/remarks/ effect on health	Reference
Oral cancer	La Vecchia et al	Tea consumption and cancer risk	BT associated with reduction of oral cancer (OR= 0.6; 95 % CI= 0.3-1.1) while BT consumption was ≥ 1 cup/day	[81]
Prostate cancer	Jian et al	Protective effect of green tea against prostate cancer: a case-control study in southeast China	Reduction in development of prostate cancer in men who drank 3 cups of GT/day (OR=0.27; 95 % CI=0.15-0.48)	[62]
.	Sonoda et al	A case-control study of diet and prostate cancer in Japan: possible protective effect of traditional Japanese diet	Consumption of GT ≥ 5 cups/day reduced risk of prostate cancer (OR=0.27; 95 % CI= 0.27–1.64)	[122]
Stomach cancer	Hoshiyama and Sasaba	Case-control study of stomach cancer and its relation to diet, cigarettes, and alcohol consumption in Saitama Prefecture, Japan	High intake of GT reduce stomach cancer (OR=0.8; 95 % CI=0.5-1.3, for highest vs. lowest level), lowest and highest consumption of GT, ≤4 cups/day ≥8 cups/day, respectively	[55]
	Kono et al	A case-control study of gastric cancer and diet in Northern Kyushu, Japan	Higher consumption of GT, reduce the risk of stomach cancer (OR=0.36; 95 % CI=0.16-0.80, for highest vs. lowest level), lowest and highest consumption level of GT ≤ 4 cups/day and ≥ 10 cups/day, respectively.	[76]

TABLE 5.4   (Continued)

Cancer Type	Authors/ researchers authors	Name of reports/ papers	Inference/remarks/ effect on health	Refer- ence
	Koizumi et al	No association between green tea and the risk of gastric cancer: pooled analysis of two prospective studies in Japan	Consumption of GT associated with stomach cancer (OR=1.19; 95 % CI 0.89-1.59, for highest vs. lowest level), lowest and highest consumption level of GT <1 cup/day and ≥5 cups/day, respectively.	[75]
.	Galanis et al	Intakes of selected foods and beverages and the incidence of gastric cancer among the Japanese residents of Hawaii: a prospective study.	Consumption of GT associated with stomach cancer (OR=1.5; 95 % CI= 0.9-2.3, for highest vs. lowest level), lowest and highest consumption of GT, 0 cups/day ≥ 2 cups/day.	[42]

OR = odds ratio; CI = 95 % confidence interval; GT=Green Tea.

harvesting time and processing methods, etc. [150]. According to study, generally 3–10 cup of tea can be consumed daily and on average one cup contains 50–100 milligram of polyphenols. Green tea doses are also recommended as cancer preventive [59].

5.7.2   EFFECTS OF GREEN TEA CONSUMPTION ON DIFFERENT TYPES OF CANCER (TABLE 5.4)

5.7.2.1   Effect on Breast Cancer

Nagata et al. [95] performed cross section study in Japan and reported that green tea consumption was associated with lowering circulating estrogen level in premenopausal women. Wu et al. [143, 144] also reported that green tea consumption was associated with lowering circulating estrogen level in postmenopausal Chinese women in Singapore and found higher consumption of green tea reduced 13% estrogen level as compared to non-drinker while consumption of black tea increases 19% level of estrogen in women. This study was performed by selecting 130 women out of which 84 were non or irregular, 27 were green tea and, 19 were black tea drinkers. Wu et al. [142] performed a case control study among Chinese, Japanese and Filipino women in Los Angeles, California, USA; and reported that there was a significant relationship between green tea consumption and reduction of breast cancer. Higher the consumption of green tea, lower the risk of breast cancer.

5.7.2.2   Effect on Esophageal Cancer

Gao et al. [43] reported that daily consumption of two or three cups of green tea reduced the risk of esophageal cancer among non-smokers and people who did not drank alcohol in Shanghai. In South America, there was 38% reduction of esophageal cancer in men and women who consumed more than 500 ml of tea per day than those who did not drink tea [19]. It was found that there was no association between drinking of normal green tea and esophageal cancer. However, esophageal cancer was increased two- to three- folds when taking very hot tea indicating higher the risk of esophageal cancer with high tea temperature [28, 67].

5.7.2.3   Effect on Lung Cancer

Cancer preventive action of green tea is due to its anti-oxidant activity, inhibition of kinases protein, inhibition of cell proliferation, and induction of apoptosis, a programmed type of cell death that differs from necrotic cell death and is known as a normal process of cell elimination [32, 37, 41, 148]. Lung tumorigenes is induced by AA-Nitrosodiethylamine (NDEA) in A/J Mice was inhibited by green tea. Another study reported by Ohno et al. [102] on Okinawan tea (partially fermented and similar to green tea) showed prevention of lung cancer especially in women, however researchers did not ask how much green tea they drink, so it was not clear that anti-carcinogenic effect were from one or both factors. Higher green tea consumption prevents the risk of lung cancer, while taking 2 cup of green tea per day can reduce 18% risk of the lung cancer [128].

5.7.2.4   Effect on Oral Cancer

La Vecchia et al. [81] reported reduction in oral cancer in non-alcohol drinkers with consuming black tea ≥1 cup/day. However, results were statistically not-significant. Green tea and the constituents present in it was able to induce apoptosis in oral carcinoma cell while EGCG of green tea suppressed the growth of oral carcinoma cell thus reduction in oral cancer [56]. Oral cancer induced by cigarette smoking was inhibited by intake of mixed tea due to its preventive action on DNA damage in oral leukoplakias [84].

5.7.2.5   Effect on Prostate Cancer

According to Jian et al. [62], consumption of green tea reduced the development of prostate cancer in men. They reported that taking of 3 cups of green tea frequently reduced prostate cancer among men however not in those who did not took green tea. Another case-control study in Japan on 140 men with prostate cancer and same numbers of hospital controls indicated that higher consumption of green tea, that is, ≥5 cups/day reduced prostate cancer in men than those who consumed <1 cup of green tea/day [122]. A report in Japan indicated a dose dependent inverse relationship among GT drinkers of Japan and risk of prostate cancer.

5.7.2.6   Effect on Stomach Cancer

Green tea shows a chemopreventive action against different types of cancer including stomach cancer. Two studies in Japan reported that higher consumption of green tea reduced onset of stomach cancer [54, 55, 76]. Hoshiyama and Sasaba [55] reported that there was a significant reduction in stomach cancer in people taking higher doses of green tea (OR = 0.8; 95% CI = 0.5–1.3, for highest vs. lowest level) while lowest and highest consumption of green tea were ≤4 cups/day and ≥8 cups/day, respectively. However, two another cohort studies in USA and Japan showed an inverse association between green tea consumption and stomach cancer, though not statically significant [42, 70, 75, 79, 105].

5.7.3   GREEN TEA AND CARDIOVASCULAR DISEASES (CVD)

High plasma cholesterol level is a big factor related to coronary heart disease can be reduced by consuming high level of green tea [45, 70, 79, 104, 109]. Asia and Japan have highest consumption of tobacco; show lowest incident of atherosclerosis and cancer due to high consumption of green tea nearly 1.2 liter per day show most beneficial effect in these regions [126]. Hypertension is a common problem in today's lifestyle and a major factor that increases high blood pressure in the arteries and leads to CVDs. High blood pressure may be defined as systolic pressure greater than 140 mm Hg and diastolic pressure greater than 90 mm Hg [151]. According to Yang et al. [151] moderate amount of consumption of green tea and oolong tea reduce hypertension.

5.7.4   MISCELLANEOUS DISEASES AND GREEN TEA

Green tea has vital impact on human and animal skins. Therefore, several skin care medicine, lotions and oils, etc. have been manufactured by using its extracts like polyphenols. Green tea leaves and buds extracts are very useful in growth of new skin cells besides minimizing the cells dying rate [83, 121]. Several studies reveal that EGCG in green tea is useful to build up the immunity against HIV and prevent the spread of this disease in primary stage [78, 98, 121]. Consumption of green tea is useful to reduce the cholesterol level in body and to decrease the chances of heart diseases up to 15 and 36%, respectively [53, 121].

5.8   SUMMARY

Green tea has been processed with different techniques for curing its natural constituents. Green tea ingredients like polyphenols, amino acids, vitamins and anti-oxidants, etc. make it as common and health promoter drink. Several eminent experts and scientists have recommended it as a refreshing drink. It is treated as health beneficial and natural beverages, contains several health curing constituents like poly phenols, vitamins, anti-oxidant properties, vitamin E and C, etc. It is frequently used for preventing diseases like cancer, brain disease, leukemia, insomnia, liver disease, diabetes and headache, etc.

The potential of green tea as therapeutic agent is of particular interest. Different species of tea and handling during processing have an impact on its health curing characteristics. Adequate amount or balanced dose of green tea works as tonic for human health but excess consumption of green tea can becomes toxic for health and several health issues can be reflected. Green tea may be effective in therapy, which is supported by a wide range of evidences provided by modern medical research. Green tea and its poly-phenol components are examples, which display cytotoxicity to cancer cells and another example is that due to its anti-microbial activity tea seems to be useful for treating certain kinds of infections.

The review of research studies in this chapter concluded that green tea has several benefits for human beings as well as controls several disease (like contribute to a reduction in the risk of CVD and some forms of cancer and other physiological functions such as anti-hypertensive effect, body weight control, anti-bacterial activity and anti-fibrotic properties, etc.). Besides prevention from diseases, green tea also nourishes the human body with vitamin C, A and minerals (like calcium, zinc fluorides, etc.). Health benefits, functional properties and nourishing characteristics have led to the inclusion of green tea in the group of beverages. Green tea also has been used for regulating body temperature and blood sugar, promoting digestion and refreshing the body. Green is recommended to ease stomach discomfort and vomiting. Green tea with a several health benefits and available at reasonable price compared to other modern beverages have made it popular in recent scientific era.

KEYWORDS

- AA-Nitrosodiethylamine
- Agricultural practices
- Alkaloids
- Amino acids
- Anthocyanidins
- Anti-aging activity
- Anti-Alzheimer activity
- Anti-angiogenic

- Anti-arthritic
- Anti-bacterial
- Anti-cancer activity
- Anti-cancerous
- Anti-carcinogenic effect
- Anti-caries activity
- Anti-diabetic activity
- Anti-inflammatory
- Anti-obesity
- Anti-oxidant activity
- Anti-oxidant compound
- Anti-oxidant potential
- Anti-oxidants
- Anti-oxidative
- Anti-Parkinson activity
- Anti-stroke activity
- Anti-tumor
- Anti-viral
- Apoptosis
- Aroma transformation
- Arteries
- Ascorbic acid oxidase
- Assamese variety
- Astringency
- Atherosclerosis
- Beneficial effect
- Beverage
- Bioactive components
- Biochemical composition
- Biochemical metabolic pathways
- Biological activities
- Bitterness
- Black tea
- Blood pressure
- Boronic acid
- Bortezomib drugs
- Breakdown of chlorophyll
- Breast cancer
- Caffeic acid
- Caffeine
- Camellia
- Camellia japonica
- Camellia sinensis
- Camellia sinensis Assamica
- Camellia sinensis Sinensis
- Cancer
- Carbohydrates
- Carcinogenesis
- Cardiovascular disease
- Carl Linnaeus
- Catalase
- Catechin
- Catechin gallate
- Cell elimination
- Cell membranes
- Cell proliferation
- Central nervous system
- Chemotherapy
- Chinese variety
- Chlorogenic acid
- Chlorophyll
- Cholesterol-lowering effects
- Chromosome structure
- Chronic disease
- Climate
- Coarse leaves
- Completely fermented
- Cultivation
- Decaffeinated tea
- Diabetes
- Diastolic pressure
- Digestive enzymes
- Disruption of cell walls
- Drum dryers

- Oral carcinoma cell
- Oral cavity
- Oral leukoplakias
- Osteoporosis
- Oxidation
- Oxidation of catechins
- Oxidation process
- Packaging machine
- Pan frying
- Pan frying process
- Parkinson's disease
- Pepsin
- Peroxidase
- Pharmaceutical industries
- Phenolic acids
- Phenolic compounds
- Phenolic hydroxyl group
- Phytochemicals
- Pigments
- Plant
- Plucking
- Plucking machine
- Polymorphism
- Polyphenol oxidase
- Polyphenol oxidation
- Polyphenols
- Popular beverage
- Processing
- Prostacyclin production
- Prostate cancer
- Proteasome inhibitors
- Protein
- Proteinkinase B
- Quercetin
- Red blood cell haemolysis
- Rolling
- Rolling process
- Rotating
- Saponins
- Secondary metabolites
- Shen Nung
- Signaling kinases
- Sleep disruption
- Slightly fermented
- Spreading
- Starch digestion
- Steaming
- Stomach cancer
- Subtropical zones
- Systolic pressure
- Tannins
- Tea production
- Tender leaves
- Thea sinensis
- Theaceae
- Theaflavins
- Therapeutic beverages
- Tray drying
- Tropical zones
- Trypsin
- Tumors
- Tyrosine kinase inhibitor
- Umami
- Unit operation
- Vacuum drying
- Vitamin-C
- Vitamin-E
- Vitamins
- Volatile compounds
- Water-soluble compounds
- Weight loss
- Withering
- Withering process
- Yabukida
- α-amylase
- α-amylase activity
- α-glucosidase

REFERENCES

1. ACOG (American College of Obstetricians and Gynecologists), (2010). ACOG Committee Opinion No. 462: Moderate caffeine consumption during pregnancy. *Obstetrics & Gynecology, 116,* (2 Pt 1), 467–468. doi:10.1097/AOG.0b013e3181eeb2a1.

2. Ahmad, N., Katiyar, S. K., & Mukhtar, H., (1998). Cancer chemopreventionby tea polyphenols. In: Ionnides, C. Ed., Nutrition and Chemical Toxicity. Wiley, West Sussex, UK, pp. 301–343.

3. An, Y., Zhang, Y., Li, C., Qian, Q., He, W., & Wang, T., (2011). Inhibitory effects of flavonoids from Abelmoschusmanihot flowers on triglycerideaccumulation in 3 T3-L1 adipocytes. *Fitoterapia, 82,* 595–600.

4. Astill, C., Birch, M. R., Dacombe, C., Humphrey, P. G., & Martin, P. T., (2001). Factors affecting the caffeine and polyphenol contents of black and green tea infusions. *Journal of Agricultural and Food Chemistry, 49*(11), 5340–5347.

5. Babu, P. V., & Liu, D., (2008). Green tea catechins and cardiovascular health: anupdate. *Current Medicinal Chemistry, 15*(18), 1840–50.

6. Balasundram, N., Sundram, K., & Samman, S., (2006). Phenolic compounds in plants and agri industrial by-products: Antioxidant activity, occurrence, and potential uses. *Food chemistry, 99,* 191–203.

7. Balentine, D. A., Wiseman, S. A., & Bouwens, L. C., (1997). The chemistry of tea flavonoids. *Critical Reviews in Food Science and Nutrition, 37*(8), 693–704.

8. Baliga, M. S., & Katiyar, S. K., (2006). Chemoprevention of photocarcinogenesis by selected dietary botanicals. *Photochemical and Photobiological Sciences, 5,* 243–253.

9. Beresford, J. N., (1989). Osteogenic stem cells and the stromal system of bone and marrow. *Clinical Orthopedics and Related Research, 240,* 270–280.

10. Bhardwaj, P., & Khanna, D., (2013). Green tea catechins: defensive role in cardiovascular disorders. *Chinese Journal of Natural Medicines, 11*(4), 0345−0353.

11. Boehm, K., Borrelli, F., Ernst, E., Habacher, G., Hung, S. K., & Milazzo, S., et al., (2009). Green tea (*Camellia sinensis*) for the prevention of cancer. *Cochrane Database of Systematic Reviews, 8* (3), 110-115.doi: 10.1002/14651858.CD005004.pub2.

12. Boyd, N. F., Rommens, J. M., Vogt, K., Lee, V. H. J. L., Yaffe, M. J., & Paterson, A. D., (2005). Mammographic breast density as an intermediate phenotype for breast cancer. *The Lancet Oncology, 6,* 798–808.

13. Braicu, C., Ladomery, M. R., Chedea, V. S., Irimie, A., & Berindan-Neagoe, I., (2013). The relationship between the structure and biological actions of green tea catechins. *Food Chemistry, 141* (3), 3282–3289.

14. Brezinova, V., (1974). Effect of caffeine on sleep: Eeg study in late middle age people. *British Journal of Clinical Pharmacology, 1,* 203–208.

15. Cabrera, C., Artacho, R., & Gimenez, R., (2006). Beneficial Effects of Green Tea—A Review. *Journal of the American College of Nutrition, 25*(2), 79–99.

16. Cai, Q., Rahn, R. O., & Zhang, R., (1997). Dietary flavonoids, quercetin, luteolin and genistein, reduce oxidative DNA damage and lipid peroxidation and quench free radicals. *Cancer Letters, 119,* 99–107.

17. Caltagirone, S., Rossi, C., & Poggi, A., (2000). Flavonoids apigenin and quercetin inhibit melanoma growth and metastatic potential. *International Journal of Cancer, 87* (4), 595–600.

18. Cano-Marquina, A., Tarín, J. J., & Cano, A., (2013). The impact of coffee on health. *Maturitas, 75*(1), 7–21.

19. Castellsague, X., Munoz, N., De Stefani, E., Victora, C. G., Castelletto, R., & Rolon, P. A., (2000). Influence of mate drinking, hot beverages and diet on esophageal cancer risk in South America. *International Journal of Cancer, 88*, 658– 664.

20. Chacko, S. M., Thambi, P. T., Kuttan, R., & Nishigaki, I., (2010). Beneficial effects of green tea: A literature review. *Chinese Medicine, 5*(13), 1–9.

21. Chakraborty, S., Kumar, S., & Basu, S., (2011). Conformational transition in the substrate binding domain of secretase exploited by NMA and its implication in inhibitor recognition: BACE1–myricetin a case study. *Neurochemistry International, 58*, 914–923.

22. Chakraborty, U., Dutta, S., & Chakraborty, B. N., (2002). Response of tea plants to water stress. *BiologiaPlantarum, 45*, 557–562.

23. Chen, L., & Zhou, Z. X., (2005). Variations of main quality componentsof tea genetic resources [*Camellia sinensis*(L.) O. Kuntze] preserved in the China National Germplasm Tea repository. *Plant Foods for Human Nutrition, 60*, 31–35.

24. Chen, Q. K, (1982). Sample Analyse of Tea Chemistry. Zhejiang Agricultural University Press, Hangzhou, China, pp. 230.

25. Chen, Z., Wang, H., You, X., & Xu, N., (2002). The chemistry of tea non-volatiles. In: Zhen, Y. Ed., Tea: Bioactivity and therapeutic potential. Taylor & Francis Inc., New York, pp. 57.

26. Cheng, O. T., (2006). All teas are not created equal The Chinese green tea and cardiovascular health. *International Journal of Cardiology, 108*, 301–308.

27. Cheng, T. O., (2004). Obesity crisis comprised of danger and opportunity. *Journal of the American Dietetic Association, 104* (10), 1546. DOI: 10.1016/j.jada.2004.08.020

28. Cheng, K. K., & Day, N. E., (1996). Nutrition and esophageal cancer. *Cancer Causes and Control, 7*, 33–40.

29. Chengelis, C. P., Kirkpatrick, J. B., Regan, K. S., Radovsky, A. E., Beck, M. J., & Morita, O., et al., (2008). 28-Day oral (gavage) toxicity studies of green tea catechins prepared for beverages in rats. *Food and Chemical Toxicology, 46*, 978–989.

30. Chiu, W., (1989). Factors affecting the production and quality of partially fermented tea in Taiwan. International Symposium on the Culture of Subtropical and Tropical Fruits and Crops.*ISHS ActaHorticulturae, 275*, 112-115. DOI: 10.17660/ActaHortic.1990.275.1

31. Choudhury, D. M. N., & Goswami, M. R., A rapid method for determination of total polyphenolic matters in tea (*Camellia sinensis* L.). *Two and a Bud 30* (1/2), 59–61.

32. Clark, J., & You, M., (2006). Chemoprevention of lung cancer by tea. *Molecular Nutrition & Food Research, 50*(2), 144–51.

33. Constable, A., Varga, N., Richoz, J., & Stadler, R. H., (1996). Antimutagenicity and catechin content of soluble instant teas. *Mutagenesis, 11*(2), 189–194.

34. Deka, A., Vita, J. A., Tea and cardiovascular disease. *Pharmacological Research*, (2011). *64*(2), 136–145.

35. Ding, M., Bhupathiraju, S. N., Satija, A., van Dam, R. M., & Hu, F. B., (2014). Long-term coffee consumption and risk of cardiovascular disease: a systematic review and a dose-response meta-analysis of prospective cohort studies. *Circulation, 129*(6), 643–659.

36. Dona, M., Dell'Aica, I., Calabrese, F., Benelli, R., Morini, M., & Albini, A., et al., (2003). Neutrophil restraint by green tea: inhibition of inflammation, associated angiogenesis, and pulmonary fibrosis. *The Journal of Immunology, 170*, 4335–4341.

37. Fesus, L., Szondy, Z., & Ura, I., (1995). Probing the molecular program of apoptosis by cancer chemopreventive agents. *Journal of Cellular Biochemistry, 22*, 151–161.

38. Figueroa, T. T. H., Rodríguez-Rodríguez, E., & Muniz, F. J. S., (2004). Green tea does a good choice for the prevention of cardiovascular disease?. *ArchivosLatino-americanos de Nutrición, 54*(4), 380–394.

39. Firenzuoli, F., Gori, L., Crupi, A., & Neri, D., (2004). Flavonoids: risks or therapeutic opportunities? *RecentiProgressi in Medicina, 95*, 345–351.

40. Fon, S. M., Yang, W. S., Gao, S., Gao, J., & Xiang, Y. B., (2011). Epidemiological studies of the association between tea drinking and primary liver cancer: a meta-analysis. *European Journal of Cancer Prevention: The Official Journal of the European Cancer Prevention Organisation (ECP) (Meta-Analysis), 20*(3), 157–165.

41. Fujiki, H., Suganuma, M., Okabe, S., Kurusu, M., Imai, K., & Nakachi, K., (2002). Involvement of TNF-alpha changes in human cancer development, prevention and palliative care. *Mechanisms of Ageing &Development, 123*(12), 1655–1663.

42. Galanis, D. J., Kolonel, L. N., Lee, J., & Nomura, A., (1998). Intakes of selected foods and beverages and the incidence of gastric cancer among the Japanese residents of Hawaii: a prospective study. *International Journal of Epidemiology, 27*, 173–180.

43. Gao, Y. T., McLaughlin, J. K., Blot, W. J., Ji, B. T., Dai, Q., & Fraumeni, J. J., (1994). Reduced risk of esophageal cancer associated with green tea consumptioa. *Journal of National Cancer Institute, 86*, 855–858.

44. Graham, H. N., (1992). Green tea composition, consumption, and polyphenol chemistry. *Preventive Medical, 21*, 334–350.

45. Grundy, S. M., (1986). Cholesterol and coronary heart disease. *A new Era of JAMA, 256*, 2849–2858.

46. Guo, Q., Zhao, B. L., Shen, S. R., Hou, J. W., Hu, J. G., & Xin, W. J., (1999). ESR study on the structure–antioxidant activity relationship of tea catechins and their epimers. *BiochimicaetBiophysicaActa – General Subjects, 1427*(1), 13–23.

47. Gupta, S., Saha, B., & Giri, A. K., (2002). Comparative antimutagenic and anticlastogenic effects of green tea and black tea: A review. *Mutation Research/Reviews in Mutation Research, 512*(1), 37–65.

48. Han, J., Britten, M., St-Gelais, D., Champagne, C. P., Fustier, P., & Salmieri, S., et al., (2011). Effect of polyphenolic ingredients on physical characteristics of cheese. *Food Research International, 44*(1), 494–497.

49. Haqqi, T. M., Anthony, D. D., Gupta, S., Ahmad, N., Lee, M. S., & Kumar, G. K., et al., (1999). Prevention of collagen-induced arthritis in mice by a polyphenolic fraction from green tea. *Proceeding of National Academic of Science USA, 96*, 4524–4529.

50. Harada, U., Chikama, A., & Saito, S., (2005). Effects of the long-term ingestion of teacatechins on energy expenditure and dietary fat oxidation in healthy subjects. *Journal of Health Science, 51*, 248–252.

51. Harbowy, M. E., & Balentine, D. A., (1997). Tea Chemistry. *Critical Review in Plant Science, 16*, 415–480.

52. He, Q., Lv, Y., & Yao, K., (2007). Effects of tea polyphenols on the activities of aamylase, pepsin, trypsin and lipase. *Food Chemistry, 101*(3), 1178–1182.

53. Hirano-Ohmori, R., Takahashi, R., & Momiyama, Y., (2005). Green tea consumption and serum malondialdehydemodified LDL concentrations in healthy subjects. *Journal of the American College of Nutrition, 24*, 342–346.

54. Hoshiyama, Y., & Sasaba, T., (1992). A case-control study of single and multiple stomach cancers in Saitama Prefecture, Japan. *Japanese Journal of Cancer Research, 83*, 937–943.

55. Hoshiyama, Y., & Sasaba, T., (1992). A Case-control study of stomach cancer and its relation to diet, cigarettes, and alcohol consumption in Saitama Prefecture, Japan. *Cancer Causes and Control, 3*, 441–448.

56. Hsu, S. D., Singh, B. B., Lewis, J. B., Borke, J. L., Dickinson, D. P., & Drake, L., (2002). Chemoprevention of oral cancer by green tea. *General Dentistry, 50*, 140–146.

57. Huang, Y. Q., Lu, X., Min, H., Wu, Q. Q., Shi, X. T., & Bian, K. Q., et al., (2015). Green tea and liver cancer risk: A meta-analysis of prospective cohort studies in Asian populations. *Nutrition, 32*(1), 3–8. doi: 10.1016/j.nut.2015.05.021

58. Huvaere, K., Nielsen, J. H., Bakman, M., Hammershoj, M., Skibsted, L. H., & Sorensen, J., et al., (2011). Antioxidant properties of green tea extract protect reduced fat soft cheese against oxidation induced by light exposure. *Journal of Agricultural and Food Chemistry, 59*(16), 8718–8723.

59. Imai, K., Suga, K., & Nakachi, K., (1997). Cancer-preventative effects of drinking tea among a Japanese population. *Preventive Medical, 26*, 769–775.

60. Jahanfar, S., & Jaafar, S. H., (2009). Effects of restricted caffeine intake by mother on fetal, neonatal and pregnancy outcome. *Cochrane Database of Systematic Reviews, 15*(2), 97–101. CD006965. Doi. 10.1002/14651858.CD006965.pub3.

61. Jia, L., & Liu, F. T., (2013). Why bortezomib cannot go with 'green'?.*Cancer Biology & Medicine, 10*(4), 206–213.

62. Jian, L., Xie, L. P., Lee, A. H., & Binns, C. W., (2004). Protective effect of green tea against prostate cancer: a case-control study in southeast China. *International Journal of Cancer, 108*, 130–135.

63. John, H. W., (1997). Tea and health: a historical perspective. *Cancer Letters, 114*(1–2), 315–317.

64. Johnson, M. K., & Loo, G., (2000). Effects of epigallocatechin gallate and quercetin on oxidative damage to cellular DNA. *Mutation Research/DNA Repair, 459*(3), 211–218.

65. Katiyar, S. K., & Mukhtar, H., (1996). Tea in chemoprevention of cancer: epidemiologic and experimental studies. *International Journal of Oncology, 8*, 221–228.

66. Katiyar, S. K., & Elmets, C. A., (2001). Green tea polyphenolic antioxidants and skin photo protection (review). *International Journal of Oncology, 18*, 1307–1313.

67. Kaufman, B. D., Liberman, I. S., & Tyshetsky, V. I., (1965). Some data concerning the incidence of oesophageal cancer in the Gurjev region of the Kazakh SSR (Russian). *VoprosyOnkologii, 11*, 78.
68. Kavanagh, K. T., Hafer, L. J., Kim, D. W., Mann, K. K., Sherr, D. H., & Rogers, A. E., et al., (2001). Green tea extracts decrease carcinogen-induced mammary tumor burden in rats and rate of breast cancer cell proliferation in culture. *Journal of Cellular Biochemistry, 82*, 387–398.
69. Khan, N., & Mukhtar, H., (2013). Tea and health: studies in humans. *Current pharmaceutical design (Literature Review), 19*(34), 6141–6147.
70. Khan, N., & Mukhtar, H., (2007). Tea polyphenols for health promotion. *Life Sciences, 81*(7), 519–533.
71. Khan, S. A., Priyamvada, S., Arivarasu, N. A., Khan, S., & Yusufi, A. N., (2007). Influence of green tea on enzymes of carbohydrate metabolism, antioxidant defense, and plasma membrane in rat tissues. *Nutrition, 23*(9), 687–695.
72. Khare, C. P., (2007). (Eds.) *Indian Medicinal Plants An Illustrated Dictionary.* Springer (India) Pvt. Ltd, Pages, 123.
73. Khokhar, S., & Magnusdottir, S. G. M., (2002). Total phenol, catechin, and caffeine-contents of teas commonly consumed in the United Kingdom. *Journal of Agricultural and Food Chemistry, 50*, 565–570.
74. Koh, L. W., Wong, L. L., Loo, Y. Y., Kasapis, S., & Huang, D., (2010). Evaluation of different teas against starch digestibility by mammalian glycosidases. *Journal of Agriculture and Food Chemistry, 58*(1), 148–154.
75. Koizumi, Y., Tsubono, Y., Nakaya, N., Nishino, Y., Shibuya, D., & Matsuoka, H., (2003). No association between green tea and the risk of gastric cancer: pooled analysis of two prospective studies in Japan. *Cancer Epidemiology, Biomarkers& Prevention, 12*, 472–473.
76. Kono, S., Ikeda, M., Tokudome, S., & Kuratsune, M., (1988). A case-control study of gastric cancer and diet in Northern Kyushu, Japan. *Japanese Journal of Cancer Research, 79*, 1067–1074.
77. Kramer, P. J., & Kozlowski, T. T., (1979). Physiology of woody plants. Academic Press. Nature, pages, 811.
78. Kuriyama, S., (2006). Green tea may do wonders for brain. *American Journal of Clinical Nutrition, 2, 83*, 355–361.
79. Kuriyama, S., Shimazu, T., Ohmori, K., Kikuchi, N., Nakaya, N., & Nishino, Y., et al., (2006). Green tea consumption and mortality due of cardiovascular disease, cancer, and all causes in japan. *Jama, 296*(10), 1255–1265.
80. Kuroda, Y., & Hara, Y., (1999). Antimutagenic and anticarcinogenicactivity of tea polyphenols. *Mutation Research, 436*, 69–97.
81. La Vecchia, C., Negri, E., Franceschi. S., D'Avanzo, B., & Boyle, P., (1992). Tea consumption and cancer risk. *Nutrition and Cancer, 17*, 27–31.
82. Langley-Evans, S. C., (2000). Consumption of black tea elicits an increase in plasma antioxidant potential in humans. *International Journal of Food Science and Nutrition, 51*(5), 309–315.
83. Lee, M. J., Maliakal, P., & Chen, L., (2002). Pharmacokinetics of tea catechins after ingestion of green tea and (-)-epigallocatechin-3-gallate by humans: formation of

different metabolites and individual variability. *Cancer Epidemiology, Biomarkers & Prevention, 11*, 1025–1032.

84. Li, N., Sun, Z., Liu, Z., & Han, C., (1998). Study on the preventive effect of tea on DNA damage of the buccal mucosa cells in oral leukoplakias induced by cigarette smoking. *Wei Sheng Yen Chiu, 27*, 173–174.

85. Lin, X., Zhang, L., Lei, H., Zhang, H., & Cheng, Y., et al., (2010). Effect of drying technologies on quality of green tea. *International Agricultural Engineering Journal, 19*(3), 30–37.

86. Liu, J., Wang, M., Peng, S., & Zhang, G., (2011). Effect of green tea catechins on the postprandial glycemic response to starches differing in amylose content. *Journal of Agriculture and Food Chemistry, 59*(9), 4582–4588.

87. Lou, L., (2002). Applications of microwave technology in the tea processing. *Fujian Tea, 1*, 23–25.

88. Matsubara, S., Rodriguez-Amaya., & Teores, B. D., (2006). De catequinas e teaflavinasemchás comercializados no Brasil.*Ciência e Tecnologia de Alimentos, 26*(2), 401–407.

89. Mbata, T. I., Debiao, L., & Saikia, A., (2008). Antibacterial activity of the crude extract of chinese green tea (*Camellia sinensis*) on listeria monocytogenes. *African Journal of Biotechnology, 7*(10), 1571–1573.

90. McKay, D. L., & Blumberg, J. B., (2002). The role of tea in human health: An update. *Journal of the American College of Nutrition, 21*, 1–13.

91. Monirul, I. M., & Han, J. H., (2012). Perceived quality and attitude toward tea & coffee by consumers. *International Journal of Business Research and Management (IJBRM), 3*(3), 100.

92. Munichiello, K. A., (2016). Green tea and iron: A poor combination?, Accessed on 05 May 2016,URL:http://worldteanews.com/tea-health-education/green-tea-and-iron-a-poor-combination

93. Murase, T., Nagasawa, A., & Suzuki, J., (2002). Beneficial effects of tea catechinsondiet-induced obesity: stimulation of lipid catabolism in the liver. *International Journal of Obesity, 26*, 1459–1464.

94. Nagalaksmi, S., (2003). Tea: An Appraisal of Processing Methods and Products. In: Hosahalli, S., Ramaswamy, G. S., Raghavan, V., Chakraverty, A., Mujumdar, A. S. Eds., Handbook of Postharvest Technology Cereals, Fruits, Vegetables, Tea, and Spices. CRC Press, pp. 741–778.

95. Nagata, C., Kabuto, M., & Shimizu, H., (1998). Association of coffee, green tea, and caffeine intakes with serum concentrations of estradiol and sex hormone-binding globulin in premenopausal Japanese women. *Nutrition and Cancer, 30*, 21–24.

96. Naghma, K., & Hasan, M., (2007). Tea polyphenols for health promotion. *Life Sciences, 81*, 519–533.

97. Naheed, Z., Barech, A. R., Sajid, M., Khan, N. A., & Hussain, R., (2007). Effect of Rolling, Fermentation and Drying on the Quality of Black tea. *Sarhad Journal of Agriculture, 23*(3), 577–580.

98. Nance, C. L., & Shearer, W. T., (2003). Is green tea good for HIV-1 infection?. *Journal of Allergy and Clinical Immunology, 112*, 851–853.

99. Narotzki, B., Reznick, A. Z., Aizenbud, D., & Levy, Y., (2012). Green tea: A promising natural product in oral health. *Archives of Oral Biology, 57*, 429–435.

100. Nehlig, A., Daval, J. L., & Debry, G., (1992). Caffeine and the central nervous system: mechanisms of action, biochemical, metabolic and psychostimulant effects. *Brain Research Review, 17*(2), 139–170.

101. Ogura, R., Ikeda, N., Yuki, K., Morita, O., Saigo, K., & Blackstock, C., et al., (2008). Genotoxy studies on green tea catechin. *Food and Chemical Toxicology, 46*(6), 2190–2220.

102. Ohno, Y., Wakai, K., Genka, K., Ohmine, K., Kawamura, T., & Tamakoshi, A., et al., (1995). Tea consumption and lung cancer risk: A case-control study in Okinawa, Japan. *Japanese Journal of Cancer Research, 86*, 1027–1034.

103. Osada, K., Takahashi, M., Hoshina, S., Nakamura, M., Nakamura, S., & Sugano, M., (2001). Tea catechins inhibit cholesterol oxidation accompanying oxidation of low density lipoprotein in vitro. *Comparative Biochemistry and Physiology Part C: Toxicology & Pharmacology, 128*(2), 153–164.

104. Pastore, R. L., & Fratellone, P., (2006). Potential health benefits of green tea (*Camellia sinensis*): A narrative review. *Explore: The Journal of Science and Healing, 2*(6), 531–539. DOI: http://dx.doi.org/10.1016/j.explore.2006.08.008

105. Pearson, D. A., Frankel, E. N., Aeshbach, R., & German, J. B., (1996). Inhibition of endothelial cell mediated oxidation of low-density lipoproteins by green tea antioxidants. *Faseb Journal, 10*, A476.

106. Qi, H., & Li, S., (2003). Dose-response meta-analysis on coffee, tea and caffeine consumption with risk of Parkinson's disease. *Geriatrics & Gerontology International, 14*, 430–439.

107. Raederstorff, D. G., Schlachter, M. F., Elste, V., & Weber, P., (2003). Effect of EGCG on lipid absorption and plasma lipid levels in rats. *The Journal of Nutritional Biochemistry, 14*, 326–332.

108. Rashidinejad, A., Birch, E. J., Sun-Waterhouse, D., & Everett, D. W., (2014). Delivery of green tea catechin and epigallocatechin gallate in liposomes incorporated into low-fat hard cheese. *Food Chemistry, 156*(1), 176–183.

109. Rice-Evans, C. A., Miller, N. J., & Paganga, G., (1997). Antioxidant properties of phenolic compounds. *Trends in Plant Science, 2*, 152–159.

110. Rietveld, A., & Wiseman, S., (2003). Antioxidant effects of tea: Evidence from human clinical trials. *Journal of Nutrition, 133*, 3285S–3292S.

111. Rio, D. D., Stewart, A. J., Mullen, W., Burns, J., Michael, E. J., & Lean, M. E. J., et al., (2004). HPLC-MS analysis of phenolic compounds an purine alkaloids in green and black tea. *Journal of Agricultural and Food Chemistry, 52*, 2807–2815.

112. Saito, S. T., Welzel, A., Suyenaga, E. S., & Bueno, F., (2006). A method for fast determination of epigallocatechingallate (EGCG), epicatechin (EC), catechin (C) and caffeine (CAF) in green tea using HPLC. *Ciencia e Tecnologia de Alimentos, 26*(2), 394–400.

113. Sandeep, P., Gopinath, C., & Mishra, M. R., (2011). Design and development of a conceptual tea leaf harvesting machine. *Sas Tech, 10* (2), 115–118.

114. Sanderson, W. G., & Grahamm, N. H., Formation of black tea aroma. *Journal of Agricultural and Food Chemistry, 21*(4), 576–585.

115. Sano, M., Tabata, M., Suzuki, M., Degawa, M., Miyase, T., & Maeda-Yamamoto, M., (2001). Simultaneous determination of twelve tea catechins by high performance liquid chromatography with electrochemical detection. *Analyst, 126*, 816–820.

116. Sartippour, M. R., Shao, Z. M., Heber, D., Beatty, P., Zhang, L., & Liu, C., et al., (2002). Green tea inhibits vascular endothelial growth factor (VEGF) induction in human breast cancer cells. *Journal of Nutrition, 132*, 2307–2311.

117. Shen, C. L., Yeh, J. K., Cao, J. J., & Wang, J. S., (2009). Green tea and bone metabolism. *Nutrition Research, 29*(7), 437–456.

118. Siddiqui, I. A., Afaq, F., Adhami, V. M., Ahmad, N., & Mukhtar, H., (2004). Antioxidants of the beverage tea in promotion of human health. *Antioxidant Redox Signal, 6*, 571–582.

119. Signorello, L. B., & McLaughlin, J. K., (2004). Maternal caffeine consumption and spontaneous abortion: a review of the epidemiologic evidence. *Epidemiology, 5*(2), 229–239.

120. Singh, V., Verma, D. K., & Singh, G., Processing technology and health benefits of green tea. *Popular Kheti, (2014, 2*(1), 23–30.

121. Sinija, V. R., & Mishra, H. N., (2008). Green tea: health benefits. *Journal of Nutritional & Environmental Medicine, 17*(4), 232–242.

122. Sonoda, T., Nagata, Y., Mori, M., Miyanaga, N., Takashima, N., & Okumura, K., et al., (2004). A case-control study of diet and prostate cancer in Japan: possible protective effect of traditional Japanese diet. *Cancer Science, 95*, 238–242.

123. Stangl, V., Dreger, H., Stangl, K., & Lorenz, M., (2007). Molecular targets of teapolyphenols in the cardiovascular system. *Cardiovascular Research, 73*(2), 348–358.

124. Sudano, R. A., Blanco, A. R., Giuliano, F., Rusciano, D., & Enea, V., (2004). Epigallocatechin-gallate enhances the activity of tetracycline in staphylococci by inhibiting its efflux from bacterial cells. *Antimicrobial Agents and Chemotherapy, 48*, 1968–1973.

125. Sueoka, N., Suganuma, M., Sueoka, E., Okabe, S., Matsuyama, S., & Imai, K., et al., (2001). A new function of green tea: Prevention of life style related diseases. *Annals of the New York Academy of Sciences, 928*, 274–280.

126. Sumpio, B. E., Cordova, A. C., Berke-Schlessel, D. W., Qin, F., & Chen, Q. H., (2006). Green tea, the "Asian Paradox," and cardiovascular disease. *Journal of the American College of Surgeons, 202*(5), 814–825.

127. Takami, S., Imai, T., Hasumura, M., Cho, Y. M., Onose, J., & Hirose, M., (2008). Evaluation of toxicity of green tea catechins with 90-day dietary administration to F344 rats. *Food and Chemical Toxicology, 46*(6), 2222–2224.

128. Tang, N., Wu, Y., Zhou, B., Wang, B., & Yu, R., (2009). Green tea, black tea consumption and risk of lung cancer: A meta-analysis. *Lung Cancer, 65*, 274–283.

129. Tijburg, L. B., Wiseman, S. A., Meijer, G. W., & Weststrate, J. A., (1997). Effects of green tea, black tea and dietary lipophilic antioxidants on LDL oxidizability and atherosclerosis in hypercholesterolaemic rabbits. *Atherosclerosis, 135*(1), 37–47.

130. Tokimitsu, I., (2004). Effects of tea catechins on lipid metabolism and body fat accumulation. *Bio Factors, 22*, 141–143.

131. Tomlins, K. I., & Mashingaidze, A., (1997). Influence of withering including leaf handling on the manufacturing and quality of black teas. A review. *Food Chemistry, 60*(4), 573–580.

132. TRISL, (2016). Tea Research Institute of Sri Lanka. Accessed on 19 May URL: http://www.tri.lk

133. Valenzuela, P. A., (2004). Tea consumption and health: Beneficial characteristics and properties of this ancient beverage. *E. Rev ChilNutr, 31*(2), 72–82.

134. Vinson, J. A., (2000). Black and green tea and heart disease: A review. *Bio Factors. 13*, 127–132.

135. Vinson, J. A., & Dabbagh, Y. A., (1998). Effect of green and black tea supplementation on lipids,lipid oxidation and fibrinogen in the hamster: mechanisms for the epidemiological benefits of tea drinking. *FEBS Letters, 433*(1–2), 44–46.

136. Wan, X. C., (2003). Tea Biochemistry. China Agricultural Press, Beijing. China. Pages, 231.

137. Wang, H., & Helliwell, K., (2000). Epimerisation of catechins in green tea infusions. *Food Chemistry, 70*(3), 337–344.

138. Wang, L., Zhang, X., Liu, J., Shen, L., & Li, Z., (2014b). Tea consumption and lung cancer risk: A meta-analysis of case-control and cohort studies. *Nutrition (Meta-Analysis), 30*(10), 1122–1127.

139. Wang, W., Yang, Y., Zhang, W., & Wu, W., (2014a). Association of tea consumption and the risk of oral cancer: a meta-analysis. *Oral Oncology (Meta-Analysis), 50*(4), 276–281.

140. Weber, J. M., Ruzindana-Umunyana, A., Imbeault, L., & Sircar, S., (2003). Inhibition of adenovirus infection and adenain by green tea catechins. *Antiviral Research, 58*, 167–173.

141. Weinreb, O., Mandel, S., Amit, T., & Youdim, M. B. H., (2004). Neurological mechanisms of green tea polyphenols in Alzheimer's and Parkinson's diseases. *The Journal of Nutritional Biochemistry*, 15, 506–516.

142. Wu, A. H., Yu, M. C., Tseng, C. C., Hankin, J., & Pike, M. C., (2003a). Green tea and risk of breast cancer in Asian Americans. *International Journal of Cancer, 106*, 574–579.

143. Wu, C. H., Lu, F. H., Chang, C. S., Chang, T. C., Wang, R. H., & Chang, C. J., (2003b). Relationship among habitual tea consumption, percent body fat, and body fat distribution. *Obesity Research, 11*, 1088–1095.

144. Wu, M. J., Weng, C. Y., Ding, H. Y., & Wu, P. J., (2005). Anti-inflammatory and antiviral effects of Glossogynetenuifolia. *Life Sciences, 76*, 1135–1146.

145. Xiao, P., & Li, Z., (2002). Botanical classification of tea plants. In: Zhen, Y. (Ed.), TEA: Bioactivity and Therapeutic Potential. Taylor & Francis Inc, New York, NY, pp. 17.

146. Xie, Y., Song, L., & Yang, X., (2006). Application of heat pump drying technology and its developmental trend. *Journal of Agricultural Mechanization Research, 4*, 12–16.

147. Xu, N., & Chen, Z., (2002). Green tea, black tea and semi-fermented tea. In: Zhen, Y. Ed., TEA: Bioactivity and Therapeutic Potential. Taylor & Francis Inc, New York, NY, pp. 35.

148. Xu, Y., Ho, C. T., Amin, S. G., Han, C., & Chung, F. L., (1992). Inhibition of tobacco-specific nitrosamine-induced lung tumorigenesis in A/J mice by green tea and its major polyphenol as antioxidants. *Cancer Research, 52*(14), 3875–3879.

149. Xua, B., & Changa, S. K. C., (2008). Effect of soaking, boiling, and steaming on total phenolic content. *Food Chemistry, 110*(3), 1–13.

150. Yamimoto, T., Juneja, L. R., Djoing-Chu, C., & Kim, M., (1997). Chemistry and applications of green tea. CRC Press, pp. 51–52

151. Yang, C. S., Hong, J., Hou, Z., & Sang, S., (2004). Green tea polyphenols: antioxidative and prooxidative effects. *Journal of Nutrition, 134*, 3181S.

152. Ye, J., Fan, F., Xu, X., & Liang, Y., (2013) Interactions of black and green tea polyphenols with whole milk. *Food Research International, 53*(1), 449–455.

153. Yilmazer-Musa, M., Griffith, A. M., Michels, A. J., Schneider, E., & Frei, B., (2012). Grape seed and tea extracts and catechin 3-gallates are potent inhibitors of aamylase and a-glucosidase activity. *Journal of Agricultural and Food Chemistry, 60*(36), 8924–8929.

154. Yuksel, Z., Avci, E., & Erdem, Y. K., (2010). Characterization of binding interactions between green tea flavanoids and milk proteins. *Food Chemistry, 121*(2), 450–456.

155. Zhang, A., Zhu, Q. Y., Luk, Y. S., Ho, K. Y., Fung, K. P., & Chen, Z. Y., (1997). Inhibitory effects of jasmine green tea epicatechin isomers on free radical-induced lysis of red blood cells. *Life Sciences, 61*(4), 383–394.

156. Zhang, L., Zhang, Z. Z., Lu, Y. N., Zhang, J. S., & Preedy, V. R., (2013). L-Theanine from green tea: Transport and effects on health. In: Preedy, R. V. Ed., Tea in Health and Disease Prevention. Academic Press, pp. 425–435.

157. Zhang, M., Holman, C. D. A. J., Huang, J. P., & Xie, X., (2007). Green tea and the prevention of breast cancer: A case control study in southeast China. *Carcinogenesis, 28*, 1074–1078.

158. Zheng, J. S., Yang, J., Fu, Y. Q., Huang, T., Huang, Y. J., & Li, D., (2013). Effects of green tea, black tea, and coffee consumption on the risk of esophageal cancer: a systematic review and meta-analysis of observational studies. *Nutrition and Cancer (Systematic Review and Meta-Analysis), 65*(1), 1–16.

159. Zhong, L., Golberg, M. S., Gao, Y. T., Hanley, J. A., Parent, M. E., & Jin, F., (2001). A population-based case-control study of lung cancer and green tea consumption among women living in Shanghai, China. *Epidemiology, 12*, 695–700.

CHAPTER 6

EFFECTS OF THERMAL PROCESSING ON NUTRITIONAL COMPOSITION OF GREEN LEAFY VEGETABLES: A REVIEW

DEEPAK KUMAR VERMA, SUDHANSHI BILLORIA,
DIPENDRA KUMAR MAHATO, AJAY KUMAR SWARNAKAR,
and PREM PRAKASH SRIVASTAV

CONTENTS

6.1   INTRODUCTION

Vegetable usually implies an edible plant or part of a plant other than a sweet fruit or seed. This usually means the leaf, stem, or root of a

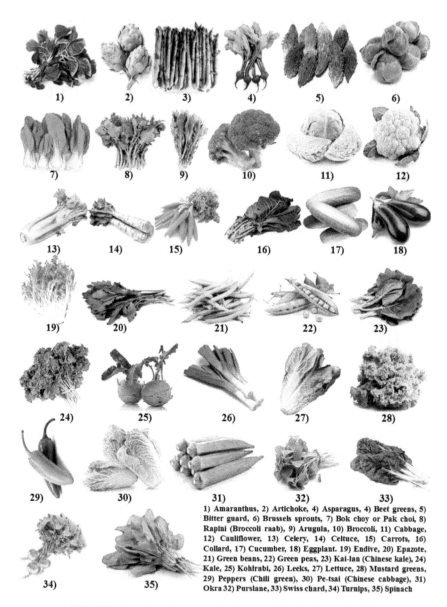

1) Amaranthus, 2) Artichoke, 4) Asparagus, 4) Beet greens, 5) Bitter guard, 6) Brussels sprouts, 7) Bok choy or Pak choi, 8) Rapini (Broccoli raab), 9) Arugula, 10) Broccoli, 11) Cabbage, 12) Cauliflower, 13) Celery, 14) Celtuce, 15) Carrots, 16) Collard, 17) Cucumber, 18) Eggplant. 19) Endive, 20) Epazote, 21) Green beans, 22) Green peas, 23) Kai-lan (Chinese kale), 24) Kale, 25) Kohlrabi, 26) Leeks, 27) Lettuce, 28) Mustard greens, 29) Peppers (Chili green), 30) Pe-tsai (Chinese cabbage), 31) Okra 32) Purslane, 33) Swiss chard, 34) Turnips, 35) Spinach

FIGURE 6.1   Some important leafy vegetables for human health.

plant. Most vegetables usually contain more than 80% of water. It has limited source of macronutrients but richest source of micronutrients [16]. There is urgent demand to include green vegetables frequently in the regular diet in a form that is easily utilized by the body, when the nutrients can be extracted from tough fiber of vegetables. Increasing the intake of green leafy vegetables (GLFs) (Table 6.1, Figure 6.1) in human diet is very important for several reasons. It contains more vitamins (viz. Vitamin A, C, and E) than other vegetables (such as carrot) and grains (such as whole wheat); provides enough minerals in the diet. They also contain quality proteins with the profile of noble amino acid (AA). Green vegetables offer important alkaline minerals such as calcium (Ca) and magnesium (Mg) that are found in insufficient amounts in fruits, nuts and seeds. Beans, broccoli, cabbage, artichoke, chard, collards, cucumber, lettuce, okra, spinach, etc. fall under green vegetables. Chlorophyll pigment is responsible for green color appearance of leafy vegetables. Several studies have shown that high consumption of GLFs in the daily diet reduces the risk of diseases and also had inverse relationship with various types of degenerative disease viz. anticarcinogenic [217], brain dysfunction [208, 224, 225], cancer [33, 40, 73, 197, 198, 216], cardiovascular [3, 149, 216], diabetes [105], heart diseases [91, 117], high blood pressure [201], inflammation [32, 168, 224], etc. The progress on "the risk of degenerative diseases" can delay and be reduced due to the presence of crucial nutrients like vitamins and minerals and additional vital constituents, which include dietary fibers and phytochemicals [18, 50, 101, 119, 134, 149, 197, 200].

Nutritional values of vegetables are numerous but these vegetables when they are exposed to processing, heat treatment and cooking, the nutritional value of these vegetables will be changed and sometimes it can also be destroyed. In this chapter, the nutrient content of various processed and treated green vegetables are discussed. In some cases, the nutrient content of processed vegetables has been reported as being greater than that of fresh (raw) vegetables. It was found that the processing and heat treatment has both positive and negative effects on the nutrient content of vegetables, which depend on different factors like altering conditions for processing and morphological and nutritional attributes of that particular species. This chapter also reviews the effects of thermal processing on nutritional components of green vegetables.

TABLE 6.1 List of Some Important Leafy Vegetables

Common Name	Scientific name	Family
Amaranthus	*Amaranthus viridis*	Amaranthaceae
Artichoke	*Cynara cardunculus var. scolymus*	Asteraceae
Asparagus	*Asparagus officinalis*	Asparagaceae
Beet greens	*Beta vulgaris*	Chenopodiaceae
Bitter guard	*Momordica charantia*	Cucurbitaceae
Brussels sprouts	*Brassica oleracea Gemmifera Group*	Brassicaceae
Bok choy or Pak choi	*Brassica rapa subsp. chinensis*	Brassicaceae
Rapini (Broccoli raab)	*Brassica rapa var. rapi*	Brassicaceae
Arugula	*Eruca sativa*	Brassicaceae
Broccoli	*Brassica oleracea Italica Group*	Brassicaceae
Cabbage	*Brassica oleracea Capitata Group*	Brassicaceae
Cauliflower	*Brassica oleracea Botrytis Group*	Brassicaceae
Celery	*Apium graveolens*	Apiaceae
Celtuce	*Lactuca sativa var. angustata*	Asteraceae
Carrots	*Daucus carota subsp. sativus*	Apiaceae
Collard	*Brassica oleracea Acephala Group*	Brassicaceae
Cucumber	*Cucumis sativus*	Cucurbitaceae
Eggplant	*Solanum melongena*	Solanaceae
Endive	*Cichorium endivia*	Asteraceae
Epazote	*Dysphania ambrosioides*	Amaranthaceae
Green beans	*Phaseolus vulgaris*	Fabaceae
Green peas	*Pisum sativum*	Fabaceae
Kai-lan (Chinese kale)	*Brassica oleracea Alboglabra Group*	Brassicaceae
Kale	*Brassica oleracea Acephala Group*	Brassicaceae
Kohlrabi	*Brassica oleracea Gongylodes Group*	Brassicaceae
Leeks	*Allium ampeloprasum*	Amaryllidaceae
Lettuce	*Lactuca sativa*	Asteraceae
Mustard greens	*Brassica juncea*	Brassicaceae

TABLE 6.1   (Continued)

Common Name	Scientific name	Family
Peppers (Chili green)	*Capsicum annuum*	Solanaceae
Pe-tsai (Chinese cabbage)	*Brassica rapa, subspecies pekinensis*	Brassicaceae
Okra	*Abelmoschus esculentus*	Malvaceae
Purslane	*Portulaca oleracea*	Portulacaceae
Swiss chard	*Beta vulgaris subsp. vulgaris*	Amaranthaceae
Turnips	*Brassica rapa subsp. rapa*	Brassicaceae
Spinach	*Spinacia oleracea*	Chenopodiaceae

6.2   NUTRITIONAL COMPOSITION AND POTENTIAL BENEFITS OF GREEN VEGETABLES

Green vegetables have almost all the important nutrients from the health point of view that are necessary for a healthy diet of human being. Some essential components include potassium (K), iron (Fe), Vitamin B_9 and certain phytochemicals or phytonutrients [58, 63]. Phytochemicals are substances that are naturally found in the vegetables [31, 58, 63, 125]. Some of the most known phytonutrients are: carotenoids, flavanoids, lutein, zeaxanthin, β-carotene, β-cryptoxanthin, etc. [56, 58]. These are important for the proper functioning of the body [53], some of which have been claimed to have antibacterial [5, 31, 106, 152, 155], anticarcinogenic [170, 217], anti-diabetes [105], antifungal [106], antioxidant [5, 55, 60, 61, 68, 74, 96, 153, 162, 163, 167, 177, 226] and antiviral [111] properties.

Vegetables are also known as the main source of fiber [187], which is classified into two classes: 1) soluble fiber and 2) insoluble fiber. A diet with high fibrous is employed for digestive tract cleansing and prevention of constipation. Vegetables offer ranges of vitamins from fat soluble (such as Vitamins A, E, and K) to water-soluble such as Vitamin B complex (Figure 6.2) and Vitamin C, which is required for

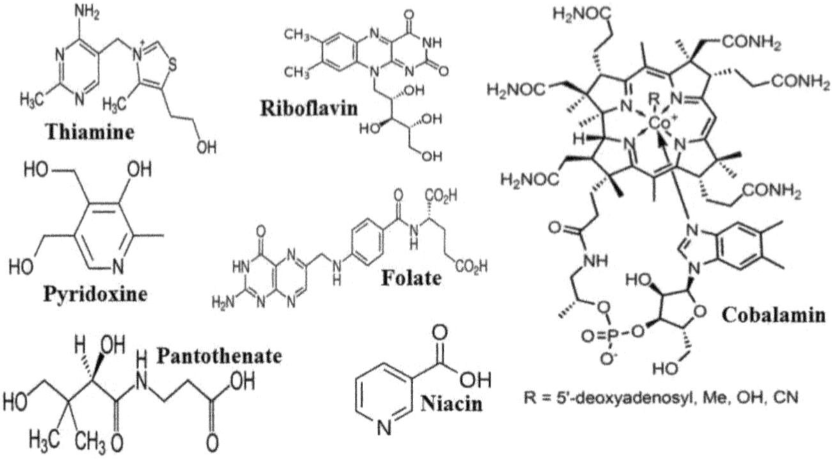

FIGURE 6.2 Structure of vitamin-B complex in green vegetables.

various physiological process of body like growth of cells, etc. Vegetables acquire minerals from the soil on which they are grown. Minerals that are mainly found in vegetables include: Fe, K, Ca, Mn, P, Mg, etc. which are essential for the overall functioning of the body. For example, the turnip (*Brassica rapa* subsp. rapa), soybeans (*Glycine max*) are rich in Vitamin A, Vitamin B_1 and Vitamin B_2, respectively. Peas (*Pisum sativum*) are rich in Vitamin A and Vitamin B_3 and broccoli is very rich in Vitamin A, Vitamin B_9 (folacin) and Vitamin C (also known as L-ascorbic acid, or simply ascorbate, chemical formula is $C_6H_8O_6$). The GLVs are also high in β-carotene (pro Vitamin A) and ascorbic acid but the yellow ones are high only in β-carotene.

Minerals that occur in tiny amounts or traces are very essential for function of the body. Green vegetables such as *Phaseolus lunatus* (lima beans, sometime also called butter beans), *G. max* and *Spinacia oleracea* (spinach) are good sources of Fe. The Ca is present in adequate quantities in turnip greens (*Brassica* spp.), soybeans (*G. max*), parsley (*Petroselinum crispum*), Chinese cabbage (*Brassica rapa*, subspecies pekinensis and chinensis), etc.

Vegetables, with the primary exception of legumes, are not especially good sources of proteins. Typically, the protein content of vegetables is only 1–2%. However, the immature seeds of legumes may be as much as 14% protein and the mature dry seeds even higher

GLVs are very beneficial and potential for human health because of having good eye sight, healthy looking skin and also to control weight gain. Carotenoids, lutein and zeaxanthin are plant pigments found in dark-GLVs, are known for its well established physiological function and protective role in eyes such as eye lens and macular region of the retina, etc. These plant pigments are important player and well-known for their protective role against both cataract and age-related macular degeneration, which are reported in the elderly as the main reason of blindness. Universally, it is also well known that vegetables have numerous nutritional value, but very difficult to say that all the nutrients are present in every vegetable. This is the main reason that all different types of vegetables must be included in one's diet as suggested by health experts. But in case of *Brassica oleracea* (broccoli), it contains almost all nutrients so consuming *B. oleracea* everyday gives higher nutrient content, vitamins and minerals than any other vegetable. *B. oleracea* is an excellent source of free-radical-scavenging Vitamin A (through its concentration of carotenoid phytonutrients), antiinflammatory Vitamin K, heart-healthy Vitamin B_9, immune-supportive Vitamin C, and digestive-health-supporting fiber [219].

Enzyme-triggering manganese (Mn); muscular-system-supportive potassium, protein, and Mg; energy-generating riboflavin, pyridoxine, and phosphorus (P); and antiinflammatory ω-3 fatty acids are categorized under well-known source of GLVs. Moreover, these are also responsible for energy-generating thiamine, niacin, pantothenic acid, and iron, bone-health enhancer Ca, and immuno-supportive zinc and Tocopherol (Vitamin E). Broccoli is well-known concentrated source of phytonutrients. Cancer-preventive attributes of broccoli is mainly obtained from glucosinolates (nutrient obtained from broccoli), which is a precursor of isothiocyanates (ITCs).

Next to nutritional effects, the pigments are necessary for the color formation in vegetables. The loss of green color is one of the main factors in quality deterioration. The most important pigments in vegetables are chlorophyll and carotenoids. The carotenoids are responsible for the yellow color of vegetables and chlorophyll is responsible for green color of vegetables. Chlorophyll is a coordination compound between Mg and the tetra-pyrrole ring [6]. The stability of the coordination complex is directly related to the stability of the green color vegetables. Chlorophyll exists as chlorophyll a and b. Chlorophyll a is the major component of the chlorophylls found in vegetables [6].

Vegetables also contain some organic acids such as citric, malic and oxalic acids are the predominant acids found in vegetables. Aside from their influence on the taste of vegetables, organic acids can have an effect on the packaging of processed products. For example, acetic and lactic acids are known to promote the pitting of tin plate in canned vegetables. Oxalic acid is aggressive toward enamel-coated plate tin plate, helping to promote the lifting of enamel from the metal. The aromatic acids and polyphenols contribute to the coloration or discoloration of canned vegetables by forming co-ordination complexes with dissolved tin or Fe. An example is the Fe-rutin complex associated with black discoloration of canned asparagus [6].

The organic acid content of green vegetables varies with maturity and cultivar. Wagner and Porter [218] reported that citric acid content of all cultivars of peas showed a rise and fall with advancing maturity. Mabesa et al. [129] showed that the cooking increases the citric acid content of peas. Lin et al. [123] reported that fumaric acid increases in concentration during storage of spinach puree. Lactic acid and succinic acid are known to increase in concentration with increasing microbial action. The green vegetables also contain aromatic acids such as benzoic, caffeic, cinnamic, chlorogenic, quinic and shikimic, etc. Baranowski and Nagel [26] have evaluated the effectiveness of hydroxycinnamic acids as anti-microbial agents.

Phytic acid content of foods is important because of the role of phytic acid and phytates in nutrition and in the processing of foods. The acid is strongly binds di- and trivalent cations like Zn, Ca, Mg, and Fe. Phytic acid is found almost exclusively in plants, with the highest concentration usually associated with seeds. The percentage of total phosphorous (P) content can be attributed to phytic acid. Because of the strong affinity of phytic acid for Ca and Zn, among other cations, the bio-availability of these minerals can be diminished by high concentrations of phytic acid in the diet. Phytic acid appears to aid in the cooking and processing of dry beans, but the nutritional quality of the total diet is improved when the phytic acid content is minimized.

6.3 THERMAL PROCESSING

Thermal processing is used primarily for sterilization of foods since the first scientific evidence witnessed in 1920 when Bigelow and Ball developed the thermal processing concept for the minimum safe sterilization process (MSSP) [20]. Thermal processing consists of heating of a food product at a temperature between 50°C to 150°C range, with primary aim to inactivate the microorganisms and endogenous enzymes. The process chosen depends on pH, microbial load, and desired shelf-life [175].

Thermal processing includes pasteurization, commercial sterilization, operations for food tenderization, and thermal pretreatments such as blanching that are conducted prior to freezing and canning to inactivate bacteria and enzymes and remove entrapped air [206]. Thermal processing operations are conventionally classified according to the intensity of heat used: pasteurization (65–85°C), sterilization (110–121°C), and ultrahigh temperature (UHT) treatment (140–160°C). Enzyme reaction rates and microbial growth increase with temperature up to a certain limit at which point inactivation begins. While thermally processed products are necessary for microbes of public health significance to be controlled, on the other hand, harsh degeneration of the quality of fresh vegetables occurs, which includes degraded color and texture application of heat to fresh vegetables, nutrient loss, cook loss, and area shrinkage [1].

In the food industries, thermal processing is well-established method for preservation of fruits, vegetables, and other food and food products that is useful for shelf-life extension. Sometimes, thermal processing (in-container sterilization, formally known as canning) is also used with high temperature thermal treatment, that is, very efficient for longtime to eliminate microbes from fruits, vegetables, and other food and food products with the concern of spoilage and public health issue [20]. The container with airtight seal maintains the inside the environment which helps to inhibit the growth and development of spoilage microbes and also help to stop recontamination and such toxins producing pathogen during the storage [20]. Furthermore, consumer prefers thermal processing of foodstuff, which results to minimum degradation of nutritive compounds [109].

6.3.1   TYPES OF THERMAL PROCESSING

Thermal processing is an important food processing technique that applies heat to food material with the vision of preservation, shelf-life extension and avoid spoilage of food, fruits, vegetables and its product. This is a single step preservation process in which heat is employed with the combinations of other preservation techniques. Employed heat treatments may vary and completely depend upon the concerned preserving action of the heat treatment and the nature and type of the food and food products.

On the basis of above discussed circumstances such as severity of heat treatment, type of heat application, purpose of heat application, etc., thermal processing operation may divided into blanching, cooking, canning, and drying. Most vegetables are commonly processed and cooked before being consumed. For most vegetables, cooking and processing is essential in transforming the product in to an edible, nutritious food [16].

6.3.1.1   Blanching

Blanching is one of the important unit operation, formally known as mild heat treatment, and is frequently employed to tissue systems prior to freezing, canning, or drying in which vegetables are heated up to the desired temperature primarily for the purpose of inactivating enzymes [54]. For example, oxidative enzymes present in vegetables which would otherwise result in undesirable changes in color, flavor, nutritive value and texture of the products [54], and similar finding also have been reported by Pilnik and Voragen [160], who found undesirable changes in flavor, nutritional value, preserving color and textural structure during processing and storage prior to freezing or drying. With removal of entrapped air and metabolic gases within vegetable cells and changes in them with water, it is easier to fill into containers for subsequent freezing or canning [160]. Blanching have also been reported in literature for removing tissues gases, shrink the product, clean and stabilize color [27]. Blanching can affect the quality of various products, viz.: excessive loss of texture, nutritional losses as shown by Arroqui et al. [15], who suggested that blanching should be applied under the suitable temperature and time conditions to avoid such negative response and minimize the disruptive textural effects. Water and steam blanching are important thermal processing, generally used in industries to heat the media for blanching. In addition to this, microwave blanching also have been studied [171].

6.3.1.1.1 Water Blanching

Water blanching can involve a low-temperature long-time (LTLT) or high-temperature short-time (HTST) process. A typical temperature ranges from 70 to 100°C depending upon the product and process conditions [124, 196] and which component, such as polyphenol oxidase (PPO), is being targeted for inactivation during the blanching process. Water blanching is performed at low temperature and results in uniform product heating but often higher leaching of minerals and vitamins [124, 196]. Some water blanchers use a screw - chain conveyer to transport the product through a blanching tank, where hot water is added; and others use a rotary drum to immerse and convey the product through the blancher.

6.3.1.1.2 Steam Blanching

Steam blanching is an alternative to water blanching. Here, product is placed on a belt conveyer that transits through a chamber containing food grade steam. It is an efficient method, because an elevated heat transfer coefficient of condensing steam is observed than that of hot water [124, 196] and is used extensively for vegetables that are cut into small pieces. Gas blanching is based on the combustion of hot gas with steam. This type of blanching has the advantage of waste reduction and nutrient retention [169].

6.3.1.1.3   Microwave Blanching

Microwave blanching can be conducted as a batch process or a continuous process. Many of the initial studies with this technology were conducted on modified home microwave ovens making comparison of unit opera-tions that would not be appropriate for an industrial use and problematic due to variability in equipment performance. Recently, the use of fiber optic temperature probes and infrared imaging make it possible to improve process control and monitoring of microwave processes allowing indus-tries to take advantage of high heat penetration and efficiencies associated with volumetric heating [54].

Microwave blanching (batch treatment at 915 MHz) of artichokes for 2 minutes completely inactivated PPO enzyme without a loss of ascorbic acid showing advantages over boiling water for 8 min and steam blanch-ing for 6 min, which resulted in 16.7 and 28.9% loss of ascorbic acid along with peroxidase inactivation [94]. There is some evidence that the quality of blanched or processed food is superior even if some peroxidase activity remains because the additional time for complete inactivation can result in browning, excessive textural softening, or changes in appearance such as ragged edges. The percentage of residual activity, which can remain without causing adverse quality changes, varies from product to product: for brussel sprouts, cauliflower, green beans and peas are 7.5–11.5%, 2.9–8.2%, 0.7–3.2% and 2–6.3%, respectively [37]. Another problem associ-ated with the complete inactivation of peroxidase is the presence of 1–10% of more heat-stable peroxidase isoenzymes in most vegetables [79, 220], which are difficult to inactivate. In some vegetables, complete inactivation of peroxidase enhances nutrients loss [37].

6.3.1.2   Cooking

The palatable attribute of the foodstuff gets improved through the thermal process known as cooking through several method of application of heat (such as, boiling, baking, broiling, roasting, frying and stewing). Though a well-known preservation technique, loss of nutrients (like vitamins and minerals) occurs due to cooking of food. However, cooking results in bioavailability of minerals (e.g., Fe) [118, 186]. Microwave cooked

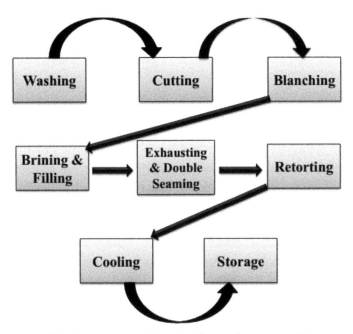

FIGURE 6.3   Process flow chart for canning of vegetables.

foodstuff causes elevated retention of Vitamin A than earth-oven cooking. Extreme losses of Vitamin C (ascorbic acid) have been reported through steam cooking. Higher retention of Vitamin B like: B_1, B_2, B_3 results from microwave cooking.

6.3.1.3   Canning

In food processing industries, canning is well-established and safe method for preservation of food if practiced properly (Figure 6.3). Fruits, vegetables, food and their products (cooked or uncooked food) are preserved by sealing them in the airtight container prior to sterilization process. Sterilization process is done in retort with heat under the high pressure. For effective sterilization, required temperature may vary with the pH of the product and is generally considered as greater than the boiling point (100°C or 212°F) of the water. This process kills microorganisms that could be a health hazard or cause the food to spoil and retain optimum quality of the preserved fruits and vegetables when at their peak of freshness. In addition, sterilization also inactivates the enzymatic activity that may the reason of undesirable changes in the color, flavor and texture [84].

6.3.1.4 Drying

Drying is a thermo-physical action having its dynamic principles governed by heat and mass transfer laws, following the mechanism of penetration of heat within the product and releasing of moisture through evaporation in a form of unsaturated gas phase [71]. In this context, safe and long storage of food products can be achieved with thus lowing of moisture content (MC). Water activity (a_w) plays an important role in spoilage of capillary-porous foodstuffs such as fruits and vegetables and thereby reduction of free water below 0.7, and growth of microbes can be hindered. Reduction of free available moisture controls biological and chemical mechanisms in fruits and vegetables [9] and increases its self-life [81, 159, 180].

Drying fundamentals involve two basic mechanisms: Constant rate period followed by falling rate period of drying [12]. Constant rate period involves in evaporation of surface water into the surrounding air (also known as drying flux period). As the relative humidity (RH) of surrounding hot air gets decreased, allowing it to absorb more moisture, it results in evaporation of food surface water or unbound water. Since rapid migration of internal water takes place for maintaining constant surface moisture, surface water remains constant at this period [89]. At this point, product temperature comes close to wet bulb temperature of drying air due to the effect of evaporative cooling [207]. During this period, driving force of drying can be measured by the difference between vapor pressure of the surface water of the food (Pw) and that of the drying air. Factors in rate of drying involve: air velocity, initial MC, relative humidity, temperature and the surface area of the food exposed to the drying air [87, 89]. The limiting factor during this step is the heat supply. Drying rate during constant rate period can be measured by the equation given by Fortes and Okos [66]:

$$dM/dt = h_T A_p (T_a - T_{wb})/L \qquad (1)$$

where, dM/dt = drying rate; h_T = convective heat transfer coefficient; A_p = surface area of the product; T_a = dry bulb temperature of dry air; T_{wb} = wet bulb temperature of air at the surface of the material; and L = heat of vaporization of water.

It has been reported that actual drying takes place at the falling rate period [81]. When drying proceeds, insufficient free moisture leads to failure of maintaining the maximum drying rate and critical MC is reached. Due to insufficient movement of bound water to product surface, rate of surface moisture loss is greater than replenished moisture or bound water [89, 207] resulting in declination of rate of drying. This point is known as falling rate period. Limiting factor at this point is case hardening [99, 100]. Factors affecting the rate of drying at this stage are collectively referred as diffusion coefficient [141]:

$$M - M_e/M_0 - M_e = e^{-kt} \tag{2}$$

where, M = moisture content; M_e = equilibrium moisture content; M_0 = initial moisture content; k = drying constant; and t = time.

The main function of drying is to remove water from the foodstuffs, which also causes some changes in color, chemical changes (affecting flavor and nutrients and shrinkage), nutritional composition [47, 48, 49, 133, 179, 180]. Since vitamins like A, C and B_1 are heat sensitive and are prone to oxidative degradation; huge loss of such vitamins has been observed during drying. According to Musa et al. [146], protein, β-carotene and Vitamin C have been reported to decrease after drying in tomatoes [146]. Joshi and Mehta [97] reported 430/100 g of leaf powder decrease in anti-nutritional factor (oxalate) in drumstick (*Moringa oleifera*) after drying. Drying has attained great importance in food industry [71, 192]. Improvement of product quality and reduction of post-harvest losses can only be achieved by the introduction of suitable drying technologies [23] like freeze drying, mechanical drying, vacuum drying, thermal drying and chemical drying [88].

6.3.1.4.1 Sun Drying

In area having favorable climatic conditions, GLFs are healthy source of nutrients whereas in semi dry-arid areas, obtainability of GLFs are limited

FIGURE 6.4   Sun drying of green vegetables with direct exposure of sun light.

[130, 138]. Therefore, to get rid of these problems, sun drying (a traditional and most ancient method) is practiced (Figure 6.4) to make availability of GLFs in the dry season [142, 180] and a cheap source of energy with no instrumental cost [107, 181]. The drawback of sun drying is its poor quality of drying and the susceptibility of food to get contaminated by dust insects and birds [11, 23, 180, 181] and unhygienic due to present of microorganisms and insects such as flies [107]. The main drawback is loss of nutrients due to direct exposure of sunlight, especially ultra-violet radiation (UVR). Trade potential and worth of the crop get affected due to loss of food quality during drying [81].

6.3.1.4.2   Mechanical Drying

Mechanical drying is a technique to remove moisture forcefully either by ambient air or through heated air by the following mechanism: (i) *Heated air drying*, which deals with the elevated temperatures resulting in rapid drying. The drying process is done till the desired final MC is reached. It uses dryers like: Batch dryer, re-circulating batch dryer, continuous flow dryer. (ii) *Low-temperature drying,* which deals with the mechanical drying, has some advantages over sun drying for various objectives such as better control over the temperature and MC, drying can be done day or night, and less labor (especially if mixing is mechanical, for example, re-circulating dryers).

6.3.1.4.3 Convective Drying

Convective drying includes drying technologies, such as: tray drying, fluidized bed, and spouted bed that provides analogous color, anthocyanin content, taste, and rehydration of halved cranberries [75]. Fluidized bed drying system includes spouted beds, which have the potential to reduce the drying time compared to tray drying and if the feature of vibration is added to it, its efficiency gets increased [75]. Grabowski et al. [75] reported perfect drying of berry at 90°C, whereas previously reported time of drying at 90°C was >50%. Drying was accelerated about 5 times in case of blueberries with spouted bed dryer (*Vaccinium corymbosum*) than tray trying at a temperature of 70°C [54]. The pre-drying of blueberries and drying at 70°C enhanced the rate of drying in spouted bed dryer and yielded the best quality drying without getting crushed or damaged than freeze and vacuum drying technologies [64].

Convective drying of onion has found its way in commercial level. For better organoleptic attributes of flakes of onion, constant drying is less effective than three stages of drying. But according to Holdsworth [92], adverse effects of convectional drying are structural changes decreasing its rehydrating attribute and excess thermal damage resulting in qualitative change of the product. Munde et al. [143] reported that 25% of onion drying time can be saved while drying is done in four-stages and is more effective than two-stage drying process. According to Lewicki [120], convective drying is a long period process and unwanted changes occur inside the product. Akbari et al. [7, 8] reported that a period of 58 min is required for drying 3 mm thick slice of onion at 76°C temperature with an air velocity of 27 m/min.

Kaymak-Ertekin and Gedik [103] concluded that 2–3 (% db) MC enhanced maximum browning rate; and browning and thiosulphinate content of the product were not significantly affected by the air velocity during drying. According to the study of Mota et al. [139], fat, ash, crude protein, fiber did not get affected by temperature of drying while sugars, acidity and Vitamin C got expressively affected. Studies on fluidized bed drying of onion concluded that it is a better technique to avoid scorching or the agglomeration of onion pieces than the tray drying, tunnel drying or convey belt drying [72, 78, 222]. Swasdisevi et al. [199] concluded that temperature of drying should be lower than 53°C, to make green color available in chopped spring onion.

6.3.1.4.4 *Freeze Drying*

According to Yang and Atallah [223], freeze drying has been reported to retain elevated amount of Vitamin C and Vitamin A and niacin, color, high rehydration rate, and low bulk density in blueberries than other drying techniques like freeze drying, convective air, vacuum oven, and microconvection. Grabowski et al. [75] reported that excellent quality of cranberries was obtained with attributes of color, taste, rehydration capacity, and anthocyanin content than with other drying technologies. Hamed and Foda [83] and Popov et al. [164] reported that best quality of dried onion was obtained in terms of color, flavor, nutrient content and rehydration attributes than traditional drying technologies. On the other hand, Andreotti et al. [14] reported that hot drying followed by freezing of onions decreases the volume and makes it friendly for storage, packaging and transportation with similar rehydrating attribute but gives a deeper color to the product. Somogyi and Luh [194] reported freeze drying as the superior drying technique with a drawback of high cost production than other drying methods. According to Freeman and Whenham [67], hot drying decreases the sensory attributes compared to freeze drying of the product.

6.3.1.4.5 *Vacuum Drying*

Vacuum drying is an excellent technique among all drying techniques, and provides low energy and capital cost for the final product [223]. Yang and

Atallah [223] reported that enhanced attributes of V. angustifolium (blueberry) were obtained than freeze drying. Grabowski et al. [75] reported that elevated attributes in terms of anthocyanin content, rehydration, color, and taste of cranberries were resulted compared to direct heating techniques. Studies revealed that for large scale production, freeze and vacuum drying is cost effective than other common drying techniques [21, 75, 223]. Mitra et al. [137] reported that in case of onions enhanced temperature and thickness effectively increased moisture diffusivity that ranged from 1.32E−10 to 1.09E−09 m2/s for untreated, and 1.32E−10 to 1.09E−01 m2/s for treated onion samples.

6.3.1.4.6   Microwave Drying

According to Beaudry et al. [28], microwave drying reduced drying time and the quality of product. Cranberries, dried under microwave, resulted in poor sensory attributes due to burn berries, blackened surface color and unpalatable flavor [28]. Kamoi et al. [98] reported comparison of microwave and freeze dried onions resulted no differences except a better rehydration in case of freeze drying. Abbasi and Azari [2] reported that microwave–vacuum–freeze drier can be recommended as a quick, simple, effectual, economic and innovative method for dehydration, which can be commercially be used.

6.3.2   NEED FOR THERMAL PROCESSING

- To render the produce edible,
- To reduce the content of plant toxins, and
- To preserve the produce, that is, to prevent spoilage due to autolysis or microbial attack.

6.3.3   THERMAL PROCESSING AND EDIBILITY

Some parts of plants are very hard to digest by our body. Therefore, cooking and processing are essential to make the food edible. The application of heat, usually cooking is the most common form of processing. It fulfills all the objectives of processing. Heat application in an aqueous medium (boiling, pressure cooking), in oil (frying) or by microwave is used so as to hydrolyze and depolymerize the hemicelluloses and the protopectins present in plant. In addition, the heat treatment (cooking) makes the cells to lose its water and leaches the nutrients, lectins and other components, so that the cell walls become permeable and soluble. In frying and microwaving, hydrolysis occur, which affects the water content of the vegetables so that the final product has a better texture, flavor and taste.

6.4   EFFECTS OF THERMAL PROCESSING ON NUTRITIONAL COMPOSITION

Nutrients are constructive wedges of the human body, which penetrate into the cell, regulating their functions and provide the energy for the continuation of their work. They can be classified into: macronutrients (proteins, fats, carbohydrates) and micronutrients (vitamins and minerals). Degradation of nutrients may occur due to processing since they are sensitive to heat, light, oxygen, pH of the solvent and/or combinations of these [85]. Degradation of nutrients may also result due to lack of handling (both household and industrial), cooking and storage [193]. It has been a common misconception that fresh vegetables are nutritionally superior to processed vegetables. Some of these misperceptions were due to erroneous comparisons made between raw produce and processed, essentially ready to eat products. Structure developing constituents (like fiber and pectic compounds), nutrients (proteins and vitamins), and non-nutritive bioactive components (like glucosinolates and polyphenols) get changed due to thermal processing, which results in variation between the sensory attributes (such as: firmness, hardness, taste, aroma, and color), and nutritive value [35, 38, 74, 122, 144, 145, 213, 226, 228]. Degradation of nutritive value and sensory attributes of foodstuff are results of thermal processing and the destruction depends on thermal resistance of the component [127]. As the objective of thermal processing is shelf-life extension and prevention of nutritive value of foodstuff, minimal processing should be done. The enhancement in nutrient content of vegetables and its effects after the processing and heat treatment are discussed under this section.

6.4.1   EFFECTS ON VITAMIN CONTENT

Vitamins are organic compounds that are very essential nutrients. They are further divided into two major classes: fat-soluble vitamins (e.g., Vitamins A, D, E, and K) and water-soluble vitamins (e.g., thiamin, riboflavin, niacin, pyridoxine, pantothenic acid, biotin, Vitamin B_{12}, choline, and ascorbic acid). All these essential nutrients can be affected by thermal processing. A wide range of researchers and research teams conducted their study on this concern and many workers have reviewed the losses of vitamins during the thermal processing of conductive heated food, fruits, vegetables and its product [128, 158, 178]. Losses during blanching and overall processing vary from vegetable to vegetable and for different vitamins. In general, higher storage temperatures tend to increase losses of thiamine, riboflavin and ascorbic acid. The vitamins viz. thiamine and riboflavin loss have been reported due to sulphur actions during pre-treatments in drying. Drying of vegetables and fruits without inactivating enzyme activity can cause carotene loss up to 80%, whereas these losses can be decreased up to 5% by adequate blanching [146].

Loss of 20–30% of Vitamin A activity may be associated with heat processing but the increased digestibility of carotenoids as a result of processing may improve the nutritional Vitamin A value of some canned vegetables [30]. GLFs (such as mint, curry, gogu and amaranth) lost 24 to 40% of β-carotene while dried under direct rays of sun [10]. A decrease in β-carotene content (un-blanched) and Vitamin C content (after blanching) in cow-pea leaves have been reported by Ndawula et al. [147]. Mulokozi and Svanberg [142] reported diminution in all-trans-α- and β-carotene (between 42 and 56%) and 9-cis-carotene (between 7 and 4 µg/g dmb) present in dried vegetables (like Mgagani, Amaranth, Cowpea, Sweet potato, Pumpkin, Ngwiba, Nsonga, Maimbe) while dried under sun compared to blanched and solar dried samples. Clydesdale et al. [51] and Thane and Reddy [205] have reported that open sun-dried milled vegetable retains 10% less all-trans-β-carotene than open sun-dried whole vegetables. sweet potatoes, cassava, cowpeas; and African spinach (Amaranthus spp.) showed retention of carotene content of about 4, 24, 42 and 8%, respectively [130]. Tanzanian vegetables showed low retention levels (ranged between 2 to 12%) after drying under sun for 24 hours according to Mosha et al. [138]. In few cases, naturally occurring antioxidants like carotenoids and Vitamin C did not get affected after drying depending on presence of phenolic compounds, place where they are located, and their binding status [41].

Vitamin C is an antioxidant and protective agent against oxidative stress. This vitamin is also very sensitive to heat and oxidation [85]. Benterud [30] reported that thiamine is the most thermo liable vitamin and about 70% of thiamin was lost during canning. Aubourg [17] reported that the groups of Vitamin B complex (viz. B_1, B_2, B_3, B_5 and B_6) are heat liable and are mostly destroyed during the sterilization process. Many workers have reported their results on vitamin losses during thermal processing in the order of 5–80%, 71–73% and 49–50% for Thiamine, Niacin and Riboflavin [25, 30, 183, 185]. Lee et al. [116] reported that peas lost less than 4% of their thiamin content during blanching, but an additional 34% loss after heat processing. Similarly, Van Buran et al. [214] reported that 90% of thiamin in snap beans was retained after blanching, but only 68% of that was present after retorting. Van Buran et al. [214] demonstrated that 58% of Vitamin B_6 in snap beans is retained during blanching and 34% after retorting. Dudek et al. [57] reported a 10–15% loss of Vitamin B_6 in water blanched bean, a 5–8% loss in steam blanched beans and 15–20% loss in water soaked retorted bean. They also reported that blanched frozen peas contained 69% of pantothenic acid of raw peas while drained canned peas contained only 30%. Folacin was retained better than pantothenic acid; 85% of the folacin was retained in drain canned peas.

Vitamins losses from vegetables are mainly due to extraction into the cooking liquid rather than their destruction [183]. Vitamin C is relatively unstable to heat, oxygen and light. The retention of Vitamin C is often used as an indication of the quality of processed foods. Processed vegetables have been reported to have a lower nutritional value such as Vitamin C loss during the processing than their respective fresh commodities [39, 114, 144, 145, 172]. If the ascorbic acid is retained well, other nutrients also will be retained well [29]. Kenny [104] estimated that the blanching step causes a loss of 25% of the ascorbic acid due to leaching of vitamin into the blanched water. Thermal processing causes additional loss of ascorbic acid due to leaching and oxidation. Benterud [30] found that only 40–75% of ascorbic acid was retained in canned green beans. Lee et al.

[116] reported that only 58% of the ascorbic acid of peas was retained after processing. It was also reported that ascorbic acid of asparagus (*Asparagus officinalis*) was greatly reduced by the cooking process [61].

Studies on Vitamin C retention of sun dried (direct and indirect exposure) vegetables have been conducted [130, 148, 189]. Negi and Roy [148] reported that sundried amaranth leaves retained elevated amount of Vitamin C compared to the vitamin content of fenugreek and savoy beet leaves and the retention depends on both the product and method of drying. Sun drying of cassava, spinach, cow pea leaves retained decreased Vitamin C than shade drying [130]. According to Maeda and Salunkhe [130], potato leaves retained maximum vitamins in shade drying and this was disagreed by Mosha et al. [138], who reported no differences in drying the above foodstuff both under shade and sun drying (provided the drying methods are season and weather dependent). The loss of Vitamin C was observed by Dewanto et al. [55] in heat-processed tomatoes at 88°C with an estimated D = 88°C value (the time taken for 90% reduction of the initial Vitamin C content at 88°C) of 276 min. The result of Dewanto et al. [55] was in the agreement with Rao et al. [172] for the kinetic study of the loss of Vitamin C in canned peas with a D = 121°C value of 246 min. Piotrowski et al. [161] reported that freeze-dried product shows elevated retention of ascorbic acid in products as compared to oven or drying in sun. Lower losses of Vitamin C have been reported under vacuum and microwave drying, since less oxidation occurs because of presence of lower levels of oxygen. Less oxidative loss of vitamins has also been reported in microwave, refractance window, low pressure superheated steam and vacuum drying.

Sablani [179] reported maximum nutrient retention in shade drying, due to the absence of light. Vitamin C losses occur primarily as the result of chemical degradation which involves an oxidation of ascorbic acid to dehydroascorbic acid (DHAA), followed by hydrolysis to 2,3-diketogulonic acid and further polymerization to form other nutritionally inactive products [77]. These chemical phenomenons are due to heat only because heat is well-known to have key role as catalyst for ascorbic acid oxidation process and thus thermal processing resulted in loss of Vitamin C content in vegetables [77]. According to Dewanto et al. [55], there were no changes in total phenolics and total flavonoids content in tomatoes with thermal treatment at 88°C. Oluwalana [157] found that simple sun drying of amaranth leaves caused decrease in Vitamin C content and elevated microbial loads compared to the blanched sundried samples, but both the values were higher than oven drying. Latapi and Barrett [112] reported huge loss of ascorbic acid in tomatoes during open sun drying. Significant loss of Vitamin C has been reported in ripe tomatoes [69, 80], whereas 54% of Vitamin C has been reported by Ojimelukwe [156].

The nutrient content of processed spinach has been reported as being greater than that of fresh spinach, that is, the carotene content of frozen and canned spinach was 60% higher than that of the fresh product [131]. It was found that blanching and processing produced relatively little change in the carotene content of peas, that is, the carotene content of processed peas was slightly higher than raw peas [116]. The total carotenoid fraction of canned carrots and spinach was higher than that of the raw vegetables. Thiamine is relatively unstable vitamin and it can be easily destroyed by heat, oxygen and pH values of 7 or higher. As a water-soluble vitamin, it can be lost during blanching and canning operations but it was found that thiamin retention in strained lima beans can be increased from approximate 60% to more than 80% by use of aseptic canning methods. It was also found that riboflavin is relatively stable during processing of asparagus, green bean, and spinach. Lee et al. [116] found better retention of Vitamin B_6 in peas after blanching. Raab et al. [166] found no loss of Vitamin B_6 of lima beans during heat processing. Benterud [30] reported that 80–100% of ascorbic acid was retained in canned asparagus, 40–75% in canned green beans and 75–95% in canned apricot. Ndawula et al. [147] reported elevated loss of β-carotene and vitamin were 58% and 84%, respectively in cowpea leaves when dried openly under sun.

6.4.2 EFFECTS ON MINERAL CONTENT

Minerals are classified into three main group: (i) macro-minerals (Ca, P, Na, Cl, and Mg); (ii) micro-minerals (Fe, Zn and Cu); and (iii) the ultra-trace minerals (Cr, F, Si, As, B, V, Ni, Cd, Li, Pb, Se, I, Mo, Mn and Co). Some of the ultra-trace minerals are very toxic, like As, Cd and Pb. Almost all of such minerals may have their toxic effect on body if consumed in large and inadequate amounts. Minerals are not destroyed by thermal processing, but losses through leaching into processing water can occur [29]. Odland and Eheart [154] reported that water blanching of broccoli reduced the P and K contents more than steam blanching. Schimitt and Weaver [182] found that chromium and zinc are lost during canning of bush beans and kale. Mineral losses were most severe in canned products that were drained before analysis. K is probably the most sensitive mineral [4], which can be lost from vegetables mainly due to the extraction into the cooking liquid reported by Schroeder [183] rather than their destruction. After drying under sun, Tamer et al. [203] found that Mg among minerals was enhanced in comparison to vacuum drying, whereas other minerals viz. K, Zn and Ca losses have been reported [203].

6.4.3 EFFECTS ON PROTEIN AND AMINO ACIDS (AAS)

Proteins are a large group of complex molecules that are polymers of AAs. The AAs are simple nitrogenous organic compounds, which serve as constructive wedges of proteins. They can be classified into essential and non-essential components based on our body needs. AAs are described as the building block of protein since its first scientific report and also responsible for the nutritional value of the food protein because of AAs distribution and availability in bioavailable form. These bioavailable forms of AAs could be modified during the processing and storage. Many reports are available that report improvement in/or loss of the nutritional and physiological properties of proteins as a result of denaturation of proteins and chemical modification of AAs [65].

Protein may react with reducing sugars, rendering them nutritionally unavailable. Typically, the protein content of vegetables is only 1–2%. However, the immature seeds of legumes may be as much as 14% protein and for the mature dry seeds even higher. Peroxides formed during lipid oxidation also react with protein and vitamins, making them biologically inactive. Threonine and lysine are the most thermo liable among the essential AAs. It has long been known that proteins inhibit the action of certain mammalian enzymes. Trypsin inhibitor is found in lima

bean, Kidney bean, peas, and potatoes [165]. Trypsin inhibitor is inactivated by heating. Piotrowski et al. [161] reported that drying along with
inactivation of trypsin inhibitor and urease (anti-nutritional factor) was
resulted due to fluidized bed drying. When whole bean was autoclaved,
98% of decreased activity was reached in 10 minutes. Many scientific
report and studies report changes in individual AAs due to heating.
Reports on several canned vegetables indicate that heating was the main
reason for changes in AA content. Tanaka and Kimura [204] reported
essential AAs losses except histidine and sulfur containing AAs. Lysine
is one among them known member of highly reactive α-amino acids
that is readily chemically modified essential AA [93]. Meredith et al.
[136] found that 35% isoleucine and 14% histidine losses were occurred
during heat processing of green turnip. There are three main possible
reasons for the losses of proteins during the canning process viz.: precooking, thermal destruction, and diffusion into the liquid. However,
during cooking, some loss of total protein lysine has been reported by
Seet and Brown [185].

6.4.4   EFFECTS ON CARBOHYDRATE AND FIBER CONTENT

Carbohydrates are known as the polyhydroxy aldehydes or ketones and
their derivatives. They are classified in three main groups: monosaccharides, oligosaccharides, and polysaccharides. The polysaccharide
substances cellulose and hemi cellulose along with pectin and lignin
contribute to the fiber content of the vegetables. It has been reported
that cooking generally increases the neutral detergent fiber (NDF), acid
detergent fiber (ADF), and cellulose content on a dry weight basis [90].
The glucose content of all vegetables except for carrot and broad bean
is decreased by cooking [132]. Anderson and Clydesdale [13] found
that rigorously boiling or retorting carrots and green beans tended to
solubilize and then destroy water-soluble pectic substances. It is pos-
tulated that sucrose is formed from the combination of glucose and
fructose during cooking. Van Buren [215] reported that cooking does
not affect pectic substances. Neutral sugars were not affected except
to increase their solubility. Hemi cellulose and cellulose fractions were

little affected, and lignin changes were small [13]. After drying under sun, losses in the total sugar and reducing sugar have been reported by Tamer et al. [203].

Borowski et al. [36] indicated that there was highest protopectin content (72.9%) in the broccoli samples cooked in a convection steam oven at 125°C with 90% steam saturation for 8 min whereas lowest protopectin content (51%) was found in the Broccoli samples boiled in water. Statistically significant differences were observed by Borowski et al. [36]. Significant differences in the content of various pectin fractions in cooked broccoli have also been reported by Wu and Chang [221], Ni et al. [151], and Christiaens et al. [46].

Thermal processing method affected quality of cooked broccoli and genetic factors [35, 122, 213, 221]. According to Wu and Chang [221], broccoli was cooked at different temperature such as 50, 60, and 70°C. They observed that the cold-water-soluble pectin content fraction was reduced and the pectin fractions content soluble in sodium hexametaphosphate and hot water was higher as compared with the raw broccoli. Studies on chemical de-esterification with high-temperature treatment revealed degradation of pectin methylesterase, which was active during the preliminary phase of heating. According to Sila et al. [190], characterization of thermo-solubility and/or β-eliminative depolymerization of pectin at elevated temperatures (>80°C) may result from enhancement of water-soluble proportion at the time of thermal processing.

6.4.5   EFFECTS ON PHYTATE CONTENT

Phytates play important role in cooking quality of legumes. The soluble phytates can be lost through leaching into cooking, soaking liquids and canning brines [95]. It has been found that the beans soaked in distilled water contained upto 1/3 less phytic acids than bean soaked in salt solutions. Tabekhia and Luh [202] showed that the canning of beans resulted in 92% decrease in the phytic acid content when compared to that of raw beans. Kon and Sanshuck [108] found that an increase in the phytic acid content of beans was co-related with a decrease in cooking time. Fanasca et al. [61] found 23% increase in total phenols in cooked asparagus. They also reported that the effect of cooking process was significant and more pronounced than the effect of cultivars.

6.4.6   EFFECTS ON PIGMENTS

Pigments found in plant are of colored substance in plant itself. Generally, pigments are considered as the chemical compounds, which absorb visible radiation between about violet (380 nm) to ruby-red (760 nm). Plant produces different types of pigments, which belong to different classes of organic compounds. They have their important function to impart color to leaves, flowers, fruits, vegetables and also have an important key role in photosynthesis, growth, and development control of plants. Chlorophyll is one among them, exists as chlorophyll a and b. Chlorophyll-a is the major component of the chlorophylls found in vegetables. It is degraded at a rate 2.5 times greater than that of chlorophyll-b. Chlorophyll may be degraded by oxidation. Schwartz and Von Elbe [184] reported that the major chlorophyll degradation product found in canned vegetables was pyropheophytin. The blanching of vegetables helps to maintain the color of green vegetables by inactivating enzymes, such as peroxidase, lipases and lipooxigenases that convert chlorophyll to pheophytins. One way to help stabilize chlorophyll in processed vegetables is to maintain an alkaline pH during blanching and to reduce the loss of Mg by maintaining a high concentration of Mg in the environment. This was the basis for the use of $MgCO_3$ (magnesium carbonate) in the Blair process for peas. Recently, this process has been updated by incorporating a Mg salt in to the organic coating of cans to be used for green vegetables.

Thermal processing significantly increased the content of bioaccessible lycopene in tomatoes observed by Dewanto et al. [55]. Enhanced *trans* and *cis*-lycopene after heating of tomato samples for 15 and 30 min was resulted due to final discharge of lycopene from cell matrix. Gartner et al. [70] and Shi and Maguer [188] found that break down of cell walls through processing of food (by cooking or grinding) may enhance the bioavailability of lycopene [70, 195]. This concludes enhanced proclamation of lycopene from the cell matrix of heat-processed tomatoes resulting in increased bio-accessibility of lycopene. Shi and Maguer [188] reported the presence of lycopene in the peripheral pericarp and the skin (covering the insoluble fiber portion) of tomatoes. The increase in *trans*-lycopene content reported by Dewanto et al. [55] was greater than the predicted values for the heat-treated samples, but the values of the raw tomatoes were within a similar range and in agreement with the studies by Shi and Maguer [188]. Nguyen and Schwartz [150] also found that lycopene was relatively very stable to persist the commercial thermal processing conditions. A reduction of lycopene content from 41 to 0.93% (dry weight) have been reported by Gupta et al. [80] in pre-treated tomatoes with 10% NaCl and also found much lower loss (17% without pre-treatment) [80]. This might be due to longer exposure of solar radiation thus degrading more amount of lycopene. Latapi and Barrett [113] reported that maximum lycopene content (>75%) after sun drying was found in tomatoes which were sulphur coated with 8% $Na_2S_2O_5$ and 8% $Na_2S_2O_5$ + 10% NaCl.

6.4.7 EFFECTS ON TOXICANTS

Any chemical may be considered as toxin/toxicant, generally metabolic products produced naturally by living organism such as animals, plants, insects, or microbes [59]. They function as poison and can injure or kill humans, animals or plants and also have evolved as defense mechanisms for the purpose of repelling or killing predators or pathogens [59, 76]. All toxins are toxicants, but all toxicants are not toxins [76]. Examples are phenolic and cyanogenetic glycosides. These toxic compounds are destroyed or leached by heat processing. The potato glycoalkaloids solanine and chaonine are typical of the toxic compounds produced by plants to defend themselves. Some phenolic compounds are also toxic such as the tannins and are bound with proteins and starches and reduce the bioavailability of these nutrients so cooking destroys large amounts of the tannins present in pulses [173]. Phytohaemagglutinins that cause clumping and destruction of red cells occur predominantly in beans. Part of this compound is also toxic to the cells of the intestinal mucosa. They are normally destroyed by soaking the beans overnight followed by thorough cooking. Glucosinolates (sulphur compounds) occur in brassica (cabbage and turnips, etc.) which can cause goiter. Udofia and Obizoba [210] reported that there was elevated reduction of anti-nutrients and food toxicants and preservation of macro and micro-nutrients in atama (some GLFs consumed in Uyo communities) throughout the year compared to shade drying.

6.4.8 EFFECTS ON PESTICIDE CONTENT

Pesticides enhance the productivity of crops by saving them from getting destroyed by pests. Residues of pesticides leftover in foodstuffs in varying amounts cause harmful effects in human health [22]. They can be used to control by means of attracting, seducing, and then destroying a variety of pathogens and pests such as insects, weeds, rats and mice, bacteria and mold [212]. Generally, word pesticide is used for the following terms: animal repellent, antimicrobial, avicide, bactericide, disinfectant (antimicrobial), fungicide, herbicide, insect growth regulator, insect repellent, insecticide, molluscicide, nematicide, piscicide, predacide, rodenticide, sanitizer and termiticide.

Thermal treatments (blanching, cooking, boiling, steaming, canning, pasteurization, etc.) efficiently reduce pesticidal residues from foodstuffs [24, 102, 110]. Peripheral surface of peel contains the maximum pesticide residues in vegetables [19]. In international trading of foodstuffs pesticide residues have become a major problem [22]. Bogner [34] showed that up to 82% of the dimethoate in French beans and cauliflower was eliminated by processing. It was also found that residues of diazinon were reduced by 77% as a result of processing. Farrow et al. [62] found that washing removed 80% of the diazinon from Chinese cabbage. Hot water blanching was able to remove 60% of the residues from spinach. The same authors reported that up to 97% of the carbaryl residues could be removed from spinach and broccoli by washing. When washing was combined with a hot water blanch, nearly all of the carbaryl was removed. Hot water blanching removed 58% DDT residue from spinach and 50% from green beans

In a study by Łozowicka and Jankowska [126], most of the pesticide residues were highly reduced after thermal processing of broccoli and tomatoes. The reduction ranged from 44% for lambda-cyhalothrin in broccoli (PF = 0.66) to 97% for boscalid in tomatoes (PF = 0.03), which was almost completely eliminated after boiling. Rasmusssen et al. [174] found that boiling did not reduce chlorpyrifos, cypermethrin, deltamethrin, diazinon, endosulfan (α-endosulfan, β-endosulfan and endosulfan sulphate), fenpropathrin, iprodione, kresoxim methyl, λ-cyhalothrin, quinalphos and vinclozoline residues.

6.4.9   DETRIMENTAL EFFECTS

It has been found that cooking and processing lead to reductions in the nutrient contents and antioxidant capacity for most vegetables [60]. Loss of various nutrients by dissolving in water may result due to direct contact of water and foodstuff. In many cases, change in availability and amount of bioactive compounds after cooking and processing occurs, resulting in compositional change of vegetables. Cooking and vegetable processing such as blanching, canning, sterilizing and drying, etc. are expected to affect the yield, composition and some nutritional antioxidants, for example, heat labile Vitamin C [96]. Heat treatment and processing also affects the texture and color of vegetables. Antioxidant activity of leafy vegetables might get affected by thermal processing due to break down of antioxidant and leaching into water [162]. Loss of heat sensitive antioxidant components (like ascorbic acid and carotenoids) may occur due to thermal processing of leafy vegetables [226].

6.4.10   EFFECTS ON PHENOLIC CONTENT

Phenolic acid also gets accumulated in vacuoles [45]. By breaking down of cellular components through thermal processing, elevated amount of phenolic compound gets released along with oxidative and hydrolytic enzymes which may result in destruction of antioxidants [45] and decrease in vegetables [44]. Normal cooking temperature results adverse effect of phenolic compound [177] in spinach, komatsuna, haruna, chingensai, cabbage and Chinese cabbage. Blanching for 15 min resulted in loss of phenolic content, in varying species of spinach [121] (http://scialert.net/fulltext/?doi=ajcn.2010.93.100&org=11-537315_ja). After drying under sun, anti-oxidant activity, total phenolic matter and HMF (Hydroxy methyl furfural) were enhanced in comparison to vacuum drying [203]. *Averrhoa bilimbi* lost phenolics and antioxidant activity when dried under sun and absorption of water deactivated the PPO activity according to Chauhan and Kapfo [43]. Mueller-Harvey [140] reported that heat produced during drying under sun enzymes and phytochemicals gets deactivated and destroyed along with loss of TPC and TAA. Whereas on contrary to this, enhancement of TPC and TAA have also been reported after drying [135, 209].

6.5 MINIMIZING NUTRITIONAL LOSSES DURING COOKING AND PROCESSING

To support the nourishment and to retain the beneficial effects of green vegetables, certain procedure has to be followed during processing. In case of broccoli, low cooking temperature in a range that includes the steaming temperature of 212°F (100°C) and with a cooking times of 5 minutes, reduces the nutritional losses at the most. To obtain the maximum nutrition and flavor in broccoli, the bottom of a steamer pot should be filled with 5 cm of water. When the water started to boil, the stem has to be cooked first followed by florets and leaves. Since the fibrous stems take longer to cook, they can be prepared separately for a few minutes before adding the florets. It was reported that stir-frying of broccoli produced some fairly positive results with respect to nutrient retention in the broccoli. The nutrients such as vitamins, minerals, phenols and glucosinolates were retained after stir-frying at the temperature range of 248°–284°F (120°–140°C) for 3.5 minutes.

About 80–100% of phenolic compound gets preserved due to decreased processing temperature [177]. Compared to conventional cooking, cooking by application of steam (short cooking time) results in preserving of antioxidants and have been reported to enhance the antioxidant activity of broccoli by 23% [177]. Enhanced antioxidant activity of about 16% has been observed in cooked green asparagus cultivars compared to the raw one [162] (http://scialert.net/fulltext/?doi=ajcn.2010.93.100&org=11-537330_ja). Singh et al. [191] reported minimum damage in leafy vegetables and maximum retention of nutrients while treated at low temperature. Feng et al. [64] reported oxidative prone components retain better quality while treated in absence of oxygen. Reduction of drying time and enhanced quality of food products are obtained in microwave drying [28]. Zhong and Lima [227] reported acceleration of vacuum drying rates of sweet potato by Ohmic heating. Leaching of antioxidants into water is a major loss, and can be avoided by minimizing the use of water, consuming the boiled water and decreasing the cooking time. Thus, appropriate methods might be sought for the processing of such vegetables to retain their antioxidant components at maximum level [44].

High fiber (cellulose and hemicellulose) content provides stringiness to the texture of green beans. Though, steaming decreases the fiber content slightly, the decrease is not significant. Since some of the nutrients found in green vegetables (e.g., beans) are particularly sensitive to light, nutrient losses occur when steaming is carried out. Among them, riboflavin is mentionable, which suffers loss during cooking due to sensitivity to heat and light. Covering the pot while cooking at 300–325°F reduces the nutritional loss.

6.6 SUMMARY

Though the thermal processing enhances the nutritional property of green vegetables, yet the nutritional losses are evident in due course. Thermal process has been reported to induce several reactions in food products that may even lead to impairment of nutritional quality of food or generation of toxic compounds like acrylamide and furan. The degree of nutritional losses depends upon the type and duration of treatment methods given to specific vegetables. On an average, the proximate composition (fat, ash, crude protein, fiber) except the moisture remains unaffected while the quality parameters like sugars, acidity and Vitamin C suffer a lot.

The sensitive quality parameters like anthocyanin content, lycopene and other heat sensitive bioactive compounds are preserved through the use of either vacuum drying or freeze drying techniques but has the drawback of high cost production than other drying methods. Microwave drying with a reduced drying time also reduce the quality of the product. The effect of heat has much influence on the Vitamin C content hence vegetables rich in Vitamin C are usually not given heat treatment in order to protect its antioxidant properties. Minerals on the other hand are least affected by thermal processing. Heating leads to inactivation of several enzymes and inhibitors in protein rich vegetables.

The vegetables like potato and radish rich in carbohydrate and fiber can be heat processed without much loss of the nutrients. Among all, the pigment usually chlorophyll is prevented from being converted to pheophytin due to the inactivation of the enzymes by blanching. Most of the toxic compounds are inactivated by the thermal processing of green vegetables along with reduction of pesticide residues that are detrimental to human health. Different processing approaches affect physicochemical properties of different green vegetables which have to be looked into individually. Increased understanding of the impact of various processing on structure, digestibility and allergenic consequence of food allergens could be applied at industrial level to develop novel processing strategies aimed at production safe and healthy foods. Therefore, consumer demands for safe and nutritious food including GLFs should be given a priority in all processing methods.

KEYWORDS

- *Abelmoschus esculentus*
- Alkaline minerals
- *Allium ampeloprasum*
- Amaranthus
- *Amaranthus viridis*
- Amino acid
- Antibacterial
- Anticarcinogenic
- Antidiabetes
- Antifungal
- Antiinflammatory
- Antioxidant
- Antiviral
- Apium graveolens
- Artichoke
- Arugula
- Ascorbate
- Ascorbic Acid
- Asparagus
- *Asparagus officinalis*
- Batch process
- Beans
- Beet greens
- *Beta vulgaris*
- *Beta vulgaris subsp. vulgaris*
- Bioavailability
- Bitter guard

- Pak choi
- Panthothenic acid
- Pantothenate
- Parsley
- Pasteurization
- Pathogen
- Peas
- Peppers
- Pesticide
- *Petroselinum crispum*
- Pe-tsai
- *Phaseolus lunatus*
- *Phaseolus vulgaris*
- Phenolic
- Phenolic acid
- Phosphorus
- Phylloquinone
- Physiological function
- Physiological process
- Phytic acid
- Phytochemicals
- Phytonutrients
- Pigment
- Pisum sativum
- Plant
- Plant pigments
- Portulaca oleracea
- Potassium
- Preservation
- Preservation techniques
- Processed vegetables
- Processing
- Protective role
- Protein
- Protein content
- Proximate composition
- Public health
- Public health issue
- Purslane
- Pyridoxine
- Pyridoxol
- Qualities
- Quality proteins
- Rapini
- Raw vegetables
- Relative humidity
- Riboflavin
- Shelf-life
- Shelf-life extension
- Sodium
- *Solanum melongena*
- Soluble fiber
- Soyabeans
- Spinach
- *Spinacia oleracea*
- Spoilage
- Spoilage microbes
- Spouted bed drying
- Steam blanching
- Sterilization
- Sterilization process
- Storage
- Sun drying
- Swiss chard
- Tetra-pyrrole ring
- Thermal pretreatments
- Thermal processing
- Thermo-physical action
- Thiamine
- Tocopherol
- Toxicants
- Toxins
- Toxins producing pathogen
- Tray drying
- Turnip
- Turnip greens

- Ultrahigh temperature treatment
- *Vaccinium corymbosum*
- Vacuum drying
- Vegetable
- Vitamin A
- Vitamin B
- Vitamin B complex
- Vitamin B_1
- Vitamin B_{12}
- Vitamin B_2
- Vitamin B_3
- Vitamin B_5
- Vitamin B_6
- Vitamin B_9

- Vitamin B_c
- Vitamin C
- Vitamin D
- Vitamin E
- Vitamin K
- Vitamin M
- Waste reduction
- Water blanching
- Water-soluble
- Water-soluble vitamin
- Zeaxanthin
- α-tocopherol
- β-carotene
- β-cryptoxanthin

- ω-3 fatty acids

REFERENCES

1. Aamir, M., Ovissipour, M., Rasco, B., Tang, J., & Sablani, S., (2014). Seasonality of the thermal kinetics of color changes in whole spinach (*Spinacia oleracea*) leaves under pasteurization conditions. *International Journal of Food Properties, 17*(9), 2012–2024.

2. Abbasi, S., & Azari, S., (2009). Novel microwave–freeze drying of onion slices. *International Journal Food Science and Technology, 44*, 974–979.

3. Ackermann, R. T., Mulrow, C. D., Ramirez, G., Gardner, C. D., Morbidoni, L., & Lawrence, V. A., (2001). Garlic shows promise for improving some cardiovascular risk factors. *Archives of Internal Medicine, 161*, 813–824.

4. Adams, C. E., & Erdman, J. W., (1988). Effects of home food preparation practices on nutrient content of foods. In: Karmas, E. and Harris, R. S. Eds., Nutritional Evaluation of Food Processing. Van Nostrand Reinhold Co, New York, pp. 557–595.

5. Adewale, A., Sinbad, O. O., & Bukoye, O. E., Phytochemical composition, antioxidant properties and antibacterial activities of five west-african green leafy Vegetables. *Canadian Journal of Pure and Applied Sciences, 7*(2), 2357–2362.

6. Ahmed, J., & Shivhare, U. S., (2012). Thermal processing of vegetables. In: Sun, D.-W. Ed., Thermal Food Processing New Technologies and Quality Issues. 2nd Ed., CRC Press, pp. 383 – 412.

7. Akbari, S. H., & Patel, N. C., (2006). Optimization of parameters for good quality dehydrated onion flakes. *Journal Food Science and Technology, 43*, 603–606.

8. Akbari, S. H., Patel, N. C., & Joshi, D. C., (2001). Studies on dehydration of onion. ASAE International Annual Meeting. Held at Sacramento, California, USA, pp. 130.

9. Akpinar, E. K., & Bicer, Y., (2004). Modelling of the drying of eggplants in thin-layers. *International Journal of Food Science and Technology, 39*(1), 1–9.

10. Aletor, V. A., & Adeogun, O. A., (1995). Nutrients and anti-nutrients components of some tropical leafy vegetables. *Journal of Food Chemistry, 54,* 375–379.

11. Al-Juamily, K. E. J., Khalifa, A. J. N., & Yassen, T. A., (2007). Testing of the performance of a fruit and vegetable solar drying system in iraq. *Desalination, 209,* 163–170

12. Alonge, A. F., & Adeboye, O. A., (2012). Drying rates of some fruits and vegetables with passive solar dryers. *International Journal of Agricultural and Biological Engineering, 5*(4), 83–90.

13. Anderson, N. E., & Clydesdale, F. M., (1980). Effects of processing on the dietary fiber content of wheat bran, pureed green beans and carrots. *Journal of Food Science, 45*(6), 1533–1537.

14. Andreotti, R., Tomasicchio, M., & Macchiavelli, L., (1981). Freeze drying of carrots and onions after partial air drying. *Ind Conserve, 54*(2), 87–91.

15. Arroqui C., Rumsey T. R., Lopez A., & Virseda P., (2001). Effect of different soluble solids in the water on the ascorbic acid losses during water blanching of potato tissue. *Journal of Food Engineering, 47,* 123–126.

16. Arthey, D., & Dennis, C., (1991). Vegetable processing. VCH publishers, New York, pp. 445.

17. Aubourg, S. P., (2001). Review: Loss of quality during the manufacture of canned fish products. *Food Science and Technology International, 7*(3), 199–215.

18. Ausman, L. M., & Mayer, J., (1999). Criteria and recommendations for vitamin C intake. *Nutrition Reviews, 57,* 222–224.

19. Awasthi, M. D., (1993). Decontamination of insecticide residues on mango by washing and peeling. *Journal of Food Science and Technology, 30,* 132–133.

20. Awuah, G. B., Ramaswamy, H. S., & Economides, A., (2007). Thermal processing and quality: Principles and overview. *Chemical Engineering and Processing, 46,* 584–602.

21. Azoubel, P. M., & Murr, F. E. X., (2003). Effect of pre-treatment on the drying kinetics of cherry tomato (*Lycopersicon esculentum* var. cerasiforme). In: Welti-Chanes, J., Velez-Ruiz, F. and Barbosa-Cánovas, G. V. Eds., Transport Phenomena in Food Processing. New York, NY: CRC Press, pp. 137–151.

22. Bajwa, U., & Sandhu, K. S., (2014). Effect of handling and processing on pesticide residues in food- a review. *Journal of Food Science and Technology, 51*(2), 201–220.

23. Bala, B. K., & Janjai, S., (2009). Solar drying of fruits, vegetables, spices, medicinal plants and fish: Developments and Potentials. *International Solar Food Processing Conference,* January 14- 16, India, pp. 180.

24. Balinova, A. M., Mla denova, R. I., & Shtereva, D. D., (2006). Effects of processing on pesticide residues in peaches intended for baby food. *Food Additives & Contaminants, 23,* 895–901.

25. Banga, J. R., Perez-Martin, R. I., Callardo, J. M., & Casares, J. J., (1991). Optimization of the thermal processing of conduction-heated canned foods: Study of several objective functions. *Journal of Food Engineering, 14,* 25–51.

26. Baranowski, J. D., & Nagel, C. W., (1982). Inhibition of pseudomonas flurisecens by hydroxycinnamic acids and their alkyl esters. *Journal of Food Science, 47*(5), 1587–1589.

27. Barrett, D. M., & Theerakulkait, C., (1995). Quality indicators in blanched, frozen, stored vegetables. *Food Technology, 49*(1), 62–65.

28. Beaudry, C., Raghavan, G. S. V., & Rennie, T. J., (2003). Microwave finish drying of osmotically dehydrated cranberries, *Drying Technology, 21*(9), 1797–1810.

29. Bender, A. E., (1966). Nutritional effects of food processing. *Journal of Food Technology, 1*, 261.

30. Benterud, A., (1977). Vitamin losses during thermal processing. In: *Høyem, T. und Kvåle, O. Eds., Physical, Chemical and Biological Changes in Food caused by Thermal Processing*. Applied Science Publishers Ltd. London, pp. 110–115.

31. Bhat, R. S., & Al-Daihan, S., (2014). Phytochemical constituents and antibacterial activity of some green leafy vegetables. *Asian Pacific Journal of Tropical Biomedicine, 4*(3), 189–193.

32. Bhupathiraju, S. N., & Tucker, K. L., (2011). Greater variety in fruit and vegetable intake is associated with lower inflammation in Puerto Rican adults. *American Journal of Clinical Nutrition, 93*, 37–46.

33. Block, G., Patterson, B., & Subar, A., (1992). Fruit, vegetables and cancer prevention-A review of the epidemiologic evidence. *Nutrition and Cancer: An International Journal, 18*, 1–29.

34. Bogner, A., (1977). Studies on partial and full elimination of pesticide residue from foods of vegetable origin by preservation and cooking. *Deutsche Lebensmittel Rundschau, 73*, 149.

35. Borowski, J., Borowska, E. J., & Szajdek, A., (2005). Wpływ warunków obróbki cieplnej brokułów (*Brassica oleracea* var. italica) na zmiany polifenoli i zdolność zmiatania rodnika DPPH. *Bromatologia i Chemia Toksykologiczna, 38*, 125–131.

36. Borowski, J., Narwojsz, J., Borowska, E. J., & Majewska, K., (2015). The effect of thermal processing on sensory properties, texture attributes, and pectic changes in broccoli. *Czech Journal of Food Science, 33*, 254–260.

37. Bottcher, H., (1975). Enzyme activity and quality of frozen vegetables. I. Remaining residual activity of peroxidase. *Nahrung, 19*, 173–177.

38. Brückner, B., Schonhof, I., Kornelson, C., & Schrödter, R., (2005). Multivariate sensory profile of broccoli and cauliflower and consumer preference. *Italian Journal of Food Science, 17*, 17–32.

39. Burge, P., & Fraile, P., (1995). Vitamin C destruction during the cooking of a potato dish. *Lebensm.-Wiss. –Technology, 28*, 506–514.

40. Byers, T., & Perry, G., (1992). Dietary carotenes, vitamin C and vitamin E as protective antioxidant in human cancers. *Annual Review of Nutrition, 12*, 139–159.

41. Capecka, E., Mareczzek, A., & Leja, M., (2005). Antioxidant activity of fresh and dry herbs of some Lamiaceae species. *Food Chemistry. 93*(2), 223–226.

42. CCOHS (Canadian Centre for Occupational Health & Safety), (2016). how do microwave ovens work?. OSH Answers Fact Sheets. Accessed on 02 June URL: http://www.ccohs.ca/oshanswers/phys_agents/microwave_ovens.html

43. Chauhan, J. B., & Kapfo, W., (2013). Effect Of traditional Sun-Drying on phenolic antioxidants of averrhoa bilimbi L. *International Journal of Applied Biology and Pharmaceutical Technology*, *4*(2), 26–34.

44. Chipurura, B., Muchuweti, M., & Manditseraa, F., (2010). Effects of thermal treatment on the phenolic content and antioxidant activity of some vegetables. *Asian Journal of Clinical Nutrition*, *2*, 93–100.

45. Chism, G. W., & Haard, N. F., (1996). Characteristics of edible plant tissues. In: Fennema, O. R. Ed., Food Chemistry. 3ʳᵈ Ed., Dekker, New York, pp. 943–1011.

46. Christiaens, S., Mbong, V. B., Van Buggenhout, S., David, C. C., Hofkens, J., & Van Loey, A., et al., (2012). Influence of processing on the pectin structure-function relationship in broccoli puree. *Innovative Food Science and Emerging Technologies*, *15*, 57–65.

47. Chua, K. J., Hawlader, M. N. A., Chou, S. K., & Ho, J. C., (2002). On the study of time varying temperature drying – effect on drying kinetics and product quality. *Drying Technology, 20*(8), 1559–1577.

48. Chua, K. J., Ho, J. C., Mujumdar, A. S., Hawlader, M. N. A., & Chou, S. K., (2000a). Convective drying of agricultural products-effect of continuous and stepwise change in drying air temperature. Paper No.29. In: Kerkhof, P. J. A. M., Coumans, W. J. and Mooiweer, G. D. Proceedings of the 12ᵗʰ International Drying Symposium. Amsterdam: Elsevier Science, pp. 230.

49. Chua, K. J., Mujumdar, A. S., Chou, S. K., Hawlader, M. N. A., & Ho, J. C., (2000b). Convective drying of banana, guava and potato pieces: effect of cyclical variations of air temperature on drying kinetics and color change. *Drying Technology, 18*, 907–936.

50. Clinton, S. K., (1998). Lycopene: Chemistry, biology, and implications for human health and disease. *Nutrition Reviews, 56*, 35–51.

51. Clydesdale, F. M., Ho, C. T., Lee, C. Y., Mondy, N. I., & Shewfelt, R. L., (1991). The effects of postharvest treatment and chemical interactions on the bioavailability of ascorbic acid, thiamine, vitamin A carotenoids and minerals. *Critical Reviews in Food Science and Nutrition, 30*, 599–638.

52. Combs, G. F. J., (2008). The vitamins: fundamental aspects in nutrition and health. 3ʳᵈ Ed., Elsevier Academic Press, Ithaca, NY, pp. 255.

53. Craig, W. J., (1999). Health-promoting properties of common herbs. *American Journal of Clinical Nutrition, 70* (3), 491–499.

54. Devece, C., Rodríguez-López, J. N., Fenoll, L. G., Tudela, J., Catalá, J. M., & Reyes, E., et al., (1999). Enzyme inactivation analysis for industrial blanching applications: Comparison of microwave, conventional, and combination heat treatments on mushroom polyphenoloxidase activity. *Journal of Agricultural and Food Chemistry*, *47*(11), 4506–4511.

55. Dewanto, V., Wu, X., Adom, K. K., & Liu, R. H., (2002). Thermal processing enhances the nutritional value of tomatoes by increasing total antioxidant activity. *Journal of Agricultural and Food Chemistry, 50*, 3010–3014.

56. Dillard, C. J., & German, J. B., (2002). Phytochemicals: nutraceuticals and human health. *Journal of the Science of Food and Agriculture, 80*, 1744–1756.

57. Dudek, J. A., Elkins, E. R., Chin, H. B., & Hagen, N. E., (1980). Investigation to determine nutrient content of selected fruits and vegetables- raw processed and prepared. National Food Processor Association, Washington, D.C, pp. 321.

58. Elias, K. M., Nelson, K. O., Simon, M., & Johnson, K., (2012). Phytochemical and antioxidant analysis of methanolic extracts of four African indigenous leafy vegetables. *Annual Review of Food Science Technology, 13*(1), 37–42.

59. Ernest, H., (2004). A Textbook of Modern Toxicology. 3rd Ed., John Wiley & Sons, Inc., Hoboken, New Jersey, pp. 262.

60. Faller, A. L. K., & Fialho, E., (2009). The antioxidant capacity and polyphenol content of organic and conventional retail vegetables after domestic cooking. *Food Research International, 42,* 210–215.

61. Fanasca, S., Rouphael, Y., Venneria, E., Azzini, E., Durazzo, A., & Maiani, G., (2009). Antioxidant properties of raw and cooked spears of green asparagus cultivars. *International Journal of Food Science and Technology, 44,* 1017–1023.

62. Farrow, R. P., & Mercher, W. A., (1969). Canning operation that reduce insecticide level in prepared food and solid food waste. *Residues Review, 29,* 73–87.

63. Fasuyi, A. O., (2006). Nutritional potentials of some tropical vegetable meals, chemical characterization and functional properties. *African Journal of Biotechnology, 5*(1), 49–53.

64. Feng, H., Tang, J., Mattinson, D. S., & Fellman, J. K., (1999). Microwave and spouted bed drying of frozen blueberries: The effect of drying and pretreatment methods on physical properties and retention of flavor volatiles. *Journal of Food Processing and Preservation, 23,* 463–479.

65. Finot. P. A., (1997). Effects of processing and storage on the nutritional value of food proteins. In: *Damodaran, S. and Parof, A. Eds., Food Proteins and Their Applications.* Marcel Dekker Inc, New York.

66. Fortes, M., & Okos, M. R., (1980). Drying Theories: Their Bases and Limitations as Applied to Foods and Grains. In: *Mujumdar, A. S. Ed., Advances in Drying.* Vol-1, Hemisphere, Washington, DC, pp. 119–154.

67. Freeman, G. G., & Whenham, R. J., (2006). Changes in onion (*Allium cepa* L.) flavor components resulting from some post-harvest processes. *Journal of the Science of Food and Agriculture, 25*(5), 499–515.

68. Gacch, R. N., Kabaliye, V. N., Dhole, N. A., & Jadhav, A. D., (2010). Antioxidant potential of selected vegetables commonly used in diet in Asian subcontinent. *Indian Journal of Natural Products and Resources, 1*(3), 306–313.

69. Gallali, Y. M., Abujnah, Y. S., & Bannani, F. K., (2000). Preservation of fruits and vegetables using solar drier: a comparative study of natural and solar drying. III. chemical analysis and sensory evaluation data of the dried samples (grapes, figs, tomatoes and onions). *Renewable Energy, 19*(1/2), 203–212.

70. Gartner, C., Stahl, W., & Sies, H., (1997). Lycopene is more bioavailable from tomato paste than from fresh tomatoes. *American Journal of Clinical Nutrition, 66,* 116–122.

71. Gavrila. C., Ghiaus, A. G., & Gruia, I., (2008). Heat and Mass Transfer in Convective Drying Processes. *Excerpt from the Proceedings of the COMSOL Conference,* Hannover, German., pp. 110.

72. Gelder, A. V., Fluidized bed process for dehydration of onion, garlic and the like. US Patent 3063848, 1962, pp. 15.

73. Giovannucci, E. A review of epidemiologic studies of tomatoes, lycopene, and prostate cancer. *Experimental Biology and Medicine, 2002, 227,* 852–859.

74. Gliszczyńska-Świgło, A., Ciska, E., Pawlak-Lemańska, K., Chmielewski, J., Borkowski, T. & Tyrakowska, B., (2006). Changes in the content of health-promoting compounds and antioxidant activity of broccoli after domestic processing. *Food Additives and Contaminants*, *23*, 1088–1098.

75. Grabowski, S., Marcotte, M., Poirier, M., Kudra, T., (2002). Drying characteristics of osmotically pretreated cranberries: Energy and quality aspects. *Drying Technology: An International Journal*, *20*(10), 1989–2004.

76. Gregory Cope, W., Leidy, R. B., & Hodgson, E., (2004). Classes of Toxicants: Use Classes. In: *Ernest, H. Ed., A Textbook of Modern Toxicology*. 3rd Ed., John Wiley & Sons, Inc., Hoboken, New Jersey, pp. 49–73.

77. Gregory, J. F. Vitamins. In: *Fennema, O. R. Ed., Food Chemistry*. 3rd Ed., Dekker, New York, 1996, pp. 531–616.

78. Gummery, C. S. A., (1977). review of common onion products. *Food Trade Review*, *47*(8), 452– 454.

79. Gunes, B., & Bayindirli, A., (1993). Peroxidase and Lipoxygenase Inactivation During Blanching of Green Beans, Green Peas and Carrots. *Lebensmittel-Wissenschaft & Technologie*, *26*(5), 406–410.

80. Gupta, R. G., Nirankar, N., (1984). Drying of tomatoes. *Journal of Food Science and Technology*, *21*, 372–376.

81. Gürlek, G., Özbalta, N., Güngör, A., (2009). Solar tunnel drying characteristics and mathematical modelling of tomato. *Journal of Thermal Science and Technology*, *29*(1), 15–23.

82. Hallstrom, B., Skjolderbrand, C., & Tragardh, C., (1988). *Heat Transfer and Food Products*. Elsevier Applied Science, London, pp. 158–242.

83. Hamed, M. G. E., & Foda, Y. H., (1966). Freeze drying of onions. *Z Lebensm Unters Forsh.*, *130*, 220.

84. Harris, L. J., *Safe Methods of Canning Vegetables*. Division of Agriculture and Natural Resources, University of California, ANR Publication 8072, 2016. Accesses on 13 May 2016, URL: http://anrcatalog.ucanr.edu/pdf/8072.pdf

85. Harris, R. S., (1988). General discussion on the stability of nutrients. In: *Karmas, E. and Harris, R. S. Eds., Nutritional Evaluation of Food Processing*. Van Nostrand Reinhold Co., New York, pp. 3–6.

86. Hasan, M. N., Akhtaruzzaman, M., & Sultan, M. J., (2013). Estimation of Vitamins B-Complex (B_2, B_3, B_5 and B_6) of Some Leafy Vegetables Indigenous to Bangladesh by HPLC Method. *Journal of Analytical Sciences, Methods and Instrumentation*, *3*, 24–29.

87. Hawlader, M. N. A.; Pera, .C. O.; Tian, M. Heat Pump Drying Under Inert Atmosphere. *Proceedings of the 14th International Drying Symposium (IDS 2004)*, August 22-25 2004, Vol.-A, Sao Paulo, Brazil, 2004, pp. 309-316.

88. Hawlader, M. N. A., Pera, C. O., & Tian, M., (2005). Influnce of Different Drying Methods on Fruits' Quality. *8th Annual IEA Heat Pump Conference*, 30th May – 2nd June 2005, Las Vegas, Nevada, United States, , pp. 150.

89. Heldman, D. R., & Haptel, R. W., (1999). *Principles of Food Processing*. Aspen Publisher Inc, New York, USA, pp. 288.

90. Heranz, J., Vidal-Valverde, C., & Rojas-Hidalgo, E., (1983). Cellelose, hemicellu-loses and lignin content of raw and cooked processed vegetables. *Journal of Food Science, 48*(1), 274–275.

91. Hertog, M. G. L., Feskens, E. J. M., Hollman, P. C. H., Katan, M. B., & Kromhout, D., (1993). Dietary antioxidant flavonoids and risk of coronary heart disease: The Zutphen Elderly Study. *Lancet, 342*(8878), 1007–1011.

92. Holdsworth, S. D., (1986). Advances in the dehydration of fruits and vegetables. In: *McCarthy, D. Ed., Concentration and Drying of Foods.* London: Elsevier, pp. 293–303.

93. Hurrel, R., & Carpenter, K., (1977). Maillard reactions in foods. In: *Hoyem, T. and Kvale, O. Eds., Physical, Chemical and Biological Changes in Food Caused by Thermal Processing.* Applied Science Publishers, London, pp. 168–184.

94. Ihl, M., Shene, C., Scheuermann, E., & Bifani, V., (1994). Correlation for pigment content through color determination using tristimulus values in a green leafy veg-etable, Swiss chard. *Journal of the Science of Food and Agriculture, 66*(4), 527–531.

95. Iyer, V., Salunkhe, D. K., Sadthe, S. K., & Rockland, L. B., (1980). Quick-cooking beans (*Phaseolus vulgaris* L.): I. Investigations on quality. *Plant Foods for Human Nutrition, 30*(1), 27–43.

96. Jimenez-Monreal, A. M., GarcIa-Diz, L., MartInez-Tome, M., Mariscal, M., & Mur-cia, M. A., (2009). Influence of Cooking Methods on Antioxidant Activity of Veg-etables. *Journal of Food Science, 74*(3), 97–103.

97. Joshi, P., & Mehta, D., (2010). Effect of dehydration on the nutritive value of drum-stick leaves. *Journal of Metabolomics and Systems Biology, 1*(1), 5–9.

98. Kamoi, I., Kikuchi, S., Matsumato, S., & Obara T., (1981). Studies on dehydration of welsh onion and carrot by microwave. *Journal of Agriculture Science (Tokyo), 20ᵗʰ Anniversary Edition,* 150–168.

99. Katekawa, M. E., & Silva, M. A., (2007). Drying rates in shrinking medium: Case study of Banana. *Brazilian Journal of Chemical Engineering, 24*(4), 561–569.

100. Katekawa, M. E., & Silva, M. A., (2007). On the influence of glass transition on shrinkage in convective drying of fruits: a case study of banana drying, *Drying Tech-nology, 25*(10), 1659–1666.

101. Kaur, C., & Kapoor, H. C., (2001). Antioxidants in fruits and vegetables—the millen-nium's health. *International Journal of Food Science and Technology, 36,* 703–725.

102. Kaushik, G., Satya, S., & Naik, S. N., (2009). Food processing a tool to pesticide residue dissipation – A review. *Food Research International, 42*(1), 26–40.

103. Kaymak-Ertekin, F., & Gedik, A., (2005). Kinetic modeling of quality deterioration in onions during drying and storage. *Journal of Food Engineering, 68,* 443–453.

104. Kenny, T. A., (1978). Nutrients are lost from fruits and vegetables between field and table. *Farm Food Research, 9,* 136.

105. Kesari, A. N., Gupta, R. K., & Watal, G., (2005). Hypoglycemic effects of *Murraya koengii* on normal and alloxan-diabetic rabbits. *Journal of Ethnopharmacology, 97,* 247–251.

106. Khan, A., & Rahman, M., (2008). Antibacterial, antifungal and cytotoxic activity of amblyone isolated from *A. campanulatus. Indian Journal of Pharmacology, 40,* 41–44.

107. Kinabo, J., Mnkeni, A., Nyaruhucha. C. N. M., & Ishengoma, J., (2004). Nutrients content of food commonly consumed in Iringa and Morogoro regions. *Proceedings of the Second Collaborative Research workshop on Food Security*, May 28–30, 2003, Morogoro. Tanzania, pp. 157.

108. Kon, S., & Sanshuck, D. W., (1981). Phytate content and its effect on cooking quality of beans. *Journal of Food Processing and Preservation*, *5*(3), 169–178.

109. Kong, F., Tang, J., Rasco, B., & Crapo, C., (2007). Kinetics of salmon quality changes during thermal processing. *Journal of Food Engineering*, *83*(4), 510–520.

110. Kumari, B., (2008). Effects of household processing on reduction of pesticide residues in vegetables. *ARPN Journal of Agricultural and Biological Science*, *3*(4), 46–51.

111. Lampe, J. W., (1999). Health effects of vegetables and fruit: assessing mechanisms of action in human experimental studies1, 2, 3. *The American Journal of Clinical Nutrition*, *70*(3), 475s–490s.

112. Latapi, G., & Barrett, D. M., (2006). Influence of pre-drying treatments on quality and safety of sun-dried tomatoes. Part I: Use of steam blanching, boiling brine blanching, and dips in salt or sodium metabisulfite. *Journal of Food Science*, *71*(1), S24–S31.

113. Latapi, G., & Barrett, D. M., (2006). Influence of pre-drying treatments on quality and safety of sun-dried tomatoes. Part II. Effects of storage on nutritional and sensory quality of sun-dried tomatoes pretreated with sulfur, sodium metbisulfite, or salt. *Journal of Food Science*, *71*(1), S32–S37.

114. Lathrop, P. J., & Leung, H. K., (1980). Rates of ascorbic acid degradation during thermal processing of canned peas. *Journal of Food Science*, *45*, 152–153.

115. Lazar, M. E., Lund, D. B., Dietrich, W. C., (1971). IQB – A New Concept in Blanching. *Food Technology*, *25*, 684–686.

116. Lee, C. Y., Massey, L. M., Jr., & Van Buren, J. P., (1982). Effects of Post-Harvest Handling and Processing on Vitamin Contents of Peas. *Journal of Food Science*, *47*(3), 961–964.

117. Lee, J. H., Jang, Y., & Kim, J. Y., (2004). Dietary habits, obesity status and Cardiovascular risk factors in Koreans. *Inter Congress Series*, *1262*, 538–541.

118. Lee, K., & Clydesdale, F. M., (1981). Effect of thermal processing on endogenous and added iron in canned spinach. *Journal of Food Science*, *46*, 1064–1067.

119. Lester, G. E., (2006). Environmental regulation of human health nutrients (ascorbic acid, carotene, and folic acid) in fruits and vegetables. *HortScience*, *41*, 59–64.

120. Lewicki, P. P., (1998). Effect of pre drying, drying and rehydration on plant tissue properties, a review. *International Journal of Food Properties*, *1*, 1–22.

121. Lima, G. P. P., Lopes, T. D. V. C., Rossetto, M. R. M., & Vianello, F., (2009). Nutritional composition, phenolic compounds, nitrate content in eatable vegetables obtained by conventional and certified organic grown culture subject to thermal treatment. *International Journal of Food Science and Technology*, *44*(6), 1118–1124.

122. Lin, C. H., & Chang, C. Y., (2005). Textural change and antioxidant properties of broccoli under different cooking treatments. *Food Chemistry*, *90*(1–2), 9–15.

123. Lin, Y. D., Clydesdale, F. M., & Francis, F. J., (1970). Organic acid profiles of thermally processed spinach puree. *Journal of Food Science*, *36*(2), 240–242.

124. Lin, Z., & Schyvens, E., (1995). Influence of blanching treatments on the texture and color of some processed vegetables and fruits. *Journal of Food Processing and Preservation, 19*, 451–465.

125. Liu, R. H., (2003). Health benefits of fruits and vegetables are from additive and synergistic combinations of phytochemicals. *American Journal of Clinical Nutrition, 78*, 517S – 520S.

126. Łozowicka, B., & Jankowska, M., (2016). Comparison of the effects of water and thermal processing on pesticide removal in selected fruits and vegetables. *Journal of Elementology, 21*(1), 99–111.

127. Lund, D. B., (1975). Effect of blanching, pasteurization and sterilization on nutrients. In: *Haris, R. S. and Karmar, E. Eds., Nutritional Evaluation of Food Processing*. AVI Publishing Co Inc., West Port, Connecticut, pp. 205–239.

128. Lund, D. B., (1977). Design of thermal process for maximizing nutrient retention. *Food Technology, 2*, 71–78.

129. Mabesa, L. B., Baldwin, R. E., & Garner, G. B., (1979). Non-volatile Organic Acid Profiles of Peas and Carrots Cooked by Microwaves. *Journal of Food Protection, 42*(5), 385–388.

130. Maeda, E. E., & Salunkhe, D. K., (1981). Retention of Ascorbic Acid and Total Carotene in Solar Dried Vegetables. *Journal of Food Science, 46*, 1288–1290.

131. Martin, S., (1977). Nutrient values of frozen vegetables as compared to fresh and canned. *Quick frozen foods (Nov.), 34*, 28.

132. Martin-Villa, C., Vidal-Valverde, C., & Rojas-Hidalgo, E., (1982). High performances liquid chromatographic determination of carbohydrates in raw and cooked vegetables. *Journal of Food Science, 47*(6), 2086–2088.

133. Mayor, L., & Sereno, A. M., (2004). Modelling shrinkage during convective drying of food materials: A review. *Journal of Food Engineering, 61*, 373–386.

134. McDermott, J. H., (2000). Antioxidant nutrients: Current dietary recommendations and research update. *Journal of the American Pharmaceutical Association, 40*, 785–799.

135. Mejia-Meza, E. I., Yáñez, J. A., Remsberg, C. M., Takemoto, J. K., Davies, N. M., & Rasco, B., Clary, C., (2010). Effect of dehydration on raspberries:polyphenol and anthocyanin retention, antioxidant capacity, and antiadipogenic activity. *Journal of Food Scienc , 75*(1), H5- H12.

136. Meredith, F. I., Gaskins, M. H., & Dull, G. G., (1974). Amino acid losses in turnip greens during handling and processing. *Journal of Food Science, 39*(4), 689–691.

137. Mitra, J., Shrivastava, S. L., & Rao, P. S., (2011). Vacuum dehydration kinetics of onion slices. *Food and Bioproducts Processing, 89*, 1–9.

138. Mosha, T. C., Pace, R. D., Adeyeye, S., Laswai, H. S., & Mtebe, K., (1997). Effect of traditional processing practices on the content of total carotenoid, β-carotene, α-carotene and vitamin A activity of selected Tanzanian vegetables. *Plant Foods for Human Nutrition, 50*, 189– 201.

139. Mota, C. L., Luciano, C., Dias, A., & Barroca, M. J., (2010). Guine, R. P. F. Convective drying of onion: kinetics and nutritional evaluation. *Food and Bioproducts Processing, 88*, 115–123.

140. Mueller-Hravey, I., (2001). Analysis of hydrolysable tannins. *Animal Feed Science and Technology, 91*, 3–20.

141. Mujumdar, A. S., (1991). Drying technologies of the future. *Drying Technology*, *9*, 325–347.

142. Mulokozi, G., & Svanberg, U., (2003). Effect of traditional open sun-drying and solar cabinet drying on carotene content and Vitamin A activity of green leafy vegetables. *Plant Foods for Human Nutrition*, *58,*1–15.4

143. Munde, A. V., Agrawal, Y. C., & Shrikande, V. J., (1988). Process development for multistage dehydration of onion flakes. *Journal of Agricultural Engineering*, *25*(1), 19–24.

144. Murcia, M. A., López-Ayerra, B., Martínez-Tomé, M., & García-Carmona, F., (2000). Effect of industrial processing on chlorophyll content of broccoli. *Journal of the Science of Food and Agriculture*, *80*, 1447–1451.

145. Murcia, M. A., Lopez-Ayerra, B., Martinez-Tome, M., Vera, A. M., & Garcia-Carmona, F., (2000). Evolution of ascorbic acid and peroxidase during industrial processing of broccoli. *Journal of the Science of Food and Agriculture*, *80*, 1882–1886.

146. Musa, J. J., Idah, P. A., & Olaleye, S. T., (2010). Effect of temperature and drying time on some nutritional quality parameters of dried tomatoes. *Assumption University: AU Journal of Technology*, *14*(1), 25–32.

147. Ndawula, J., Kabasa, J. D., & Byaruhanga, Y. B., (2004). Alterations in fruit and vegetable beta-carotene and vitamin C content caused by open-sun drying, visqueen-covered and polyethylene-covered solar-dryers. *African Health Sciences*, *4*(2)125–130.

148. Negi, P. S., & Roy, S. K., (2000). Effect of blanching and drying methods on β-carotene, ascorbic acid and chlorophyll retention of leafy vegetables. *Food Science and Technology*, *33*, 295–298.

149. Ness, A., & Powles, J. W., (1997). Fruit and vegetables, and cardiovascular disease: A review. *International Journal of Epidemiology*, *26*, 1–13.

150. Nguyen, M. L., & Schwartz, S. J., (1998). Lycopene stability during food processing. *Proceedings of The Society for Experimental Biology and Medicine*, *218*, 101–105.

151. Ni, L.; Lin, D., & Barrett, D. M., (2005). Pectin methylesterase catalyzed firming effects on low temperature blanched vegetables. *Journal of Food Engineering*, *70*, 546–556.

152. Obi, K. R., & Nwanebu, F. C., (2009). Antibacterial qualities and phytochemical screening of the oil extract of cucurbita pepo. *Journal Medicinal Plant Research*, *3*(5), 429–432.

153. Oboh, G. O., Nwanna, E. E., & Elusiyan, C. A., (2006). Antioxidant and antimicrobial properties of *Telfairia occidentalis* (Fluted pumpkin) leaf extracts. *Journal of Pharmacology and Toxicology*, *1*, 167–175.

154. Odland, D., & Eheart, M. S., (1975). Ascorbic acid mineral and quality retention in frozen broccoli blanched in water steam and ammonia-steam. *Journal of Food Science*, *40*(5), 1004–1007.

155. Odoemena, C. S., & Essien, J. P., (1995). Antibacterial activity of the root extract of *T. occidentalis* (fluted pumpkin). *West African Journal of Biology and Applied Chemistry*, *40*, 1–4.

156. Ojimelukwe, C. P., (1994). Effects of processing methods on ascorbic acid retention and sensory characteristics of tomato products. *Journal of Food Science and Technology*, *31*(3), 247–248.

157. Oluwalana, I. B., (2013). Effect of processing methods on the nutritional value of *Amaranthus hybridus* L. *International Journal of Agriculture and Food Security*, *4*(15), 523–529.

158. Paulas K., (1989). Vitamin degradation during food processing and how to prevent it. In: *Somogyi, J. C. and Muller, H. R. Eds., Nutritional Impact of Food Processing.* Bibliotheca Nutrio et Dieta Karger, Basle, Switzerland, pp. 173–187.

159. Perumal, R., (2007). *Comparative performance of solar cabinet, vacuum assisted solar and open sun drying methods.* M.Sc. Dissertation, Department of Bioresource Engineering, McGill University, Montreal, Canada, pp. 110.

160. Pilnik, W., & Voragen, A. G. J., (1991). The significance of endogenous and exogenous pectic enzymes in fruit and vegetable processing. *In: Fox, P. F. Ed., Food Enzymology.* Essex, England: Elsevier Applied Science, pp. 318.

161. Piotrowski, D., Lenart, A., & Wardzynski, A., (2004). Influence of osmotic dehydration on microwave-convective drying of frozen strawberries. *Journal of Food Engineering*, *65*, 519–525.

162. Podsedek, A., (2007). Natural antioxidants and antioxidant capacity of Brassica vegetables: A review. *LWT - Food Science and Technology*, *40*(1), 1–11

163. Podsędek, A., Sosnowska, D., Redzynia, M., & Koziołkiewicz, M., (2008). Effect of domestic cooking on the red cabbage hydrophilic antioxidants. *International Journal of Food Science & Technology*, *43*, 1770–1777.

164. Popov, O. A., Efron, B. G., & Milakova, E., (1976). *Vse-soyuznyi Nauchno- issledovatel'shii Institut Konservnoii Ovoshchesushill'noi Promyshlenn-ost*, *24*, 43–50.

165. Pressey, R., (1972). Natural Enzyme inhibitors in plant tissues. *Journal of Food Science*, *37*(4), 521–523.

166. Raab, C. A., Luh, B. S., & Schweiger, B. S., (1973). Effects of heat processing on the retention of vitamin B6 in canned lima beans. *Journal of Food Science*, *38*(3), 544–545.

167. Raghavedran, M., Reddy, A. M., Yadav, P. R., Raju, A. S., & MKumar, L. S., & Comparative studies on the in vitro antioxidant properties of methanolic leafy extracts from six edible leafy vegetables of India. *Asian Journal of Pharmaceutical and Clinical Research*, *6*(3), 96–99.

168. Rahman, I., & Adcock, I. M., (2006). Oxidative stress and redox regulation of lung inflammation in COPD. *European Respiratory Journal*, *28*, 219–242.

169. Rahman, M. S., & Perera, C., (1999). Drying and food preservation. In: *Rahman, M. S. Ed., Handbook of Food Preservation.* Marcel Dekker, New York, NY, USA, pp. 192–194.

170. Rajeshkumar, N. V., Joy, K. L., Kuttan, G., Ramsewak, R. S., Nair, M. G., & Kuttan, R., (2002). Antitumour and Anticarcinogenic Activity of *Phyllanthus amarus* Extract. *Journal of Ethnopharmacology*, *81*, 17–22.

171. Ramesh, M. N., Wolf, W., Tevini, D., & Bognar, A., (2002). Microwave blanching of vegetables. *Journal of Food Science*, *67* (1), 390–398.

172. Rao, M. A., Lee, C. Y., Katz, J., & Cooley, H. J., (1981). A kinetic study of the loss of vitamin C, color, and firmness during thermal processing of canned peas. *Journal of Food Science*, *46*, 636–637.

173. Rao, P. U., & Deosthale, Y. G., (1982). Tannin content of pulse: Varietal difference and effects of germination and cooking. *Journal of the Science of Food and Agriculture, 33*(10), 1013–1016.

174. Rasmusssen, R. R., Poulsen, M. E., & Hansen H. C. B., (2003). Distribution of multiple pesticide residues in apple segments after home processing. *Food Additives & Contaminants, 20*, 1044–1063.

175. Rein, M., (2005). Co-pigmentation Reactions and Color Stability of Berry Anthocyanins. Department of Applied Chemistry and Microbiology, University of Helsinki, Helsinki, Finland, pp. 40.

176. Rickman, J. C., Barrett, D. M., & Bruhn, C. M., (2007). Nutritional comparison of fresh, frozen and canned fruits and vegetables. Part 1. Vitamins C and B and phenolic compounds. *Journal of the Science of Food and Agriculture, 87*(6), 930–994.

177. Roy, M. K., Takenaka, M., Isobe, S., & Tsushida, T., (2007). Antioxidant potential, anti- proliferative activities, and phenolic content in water-soluble fractions of some commonly consumed vegetables: Effects of thermal treatment. *Food Chemistry, 103*, 106–114.

178. Ryley, J., Abdel-Kader, Z. M., Guevara, L. V., Lappo, B. P., Lamb, J., & Mana, F., et al., (1990). The prediction of vitamin retention in food processing. *Journal of Micronutrient Analysis, 7*, 329–347.

179. Sablani, S. S., (2006). Drying of Fruits and Vegetables: Retention of Nutritional/ Functional Quality. *Drying Technology, 24*, 428–432.

180. Sagar, V. R., & Suresh, K. P., (2010). Recent advances in drying and dehydration of fruits and vegetables: a review. *Journal of Food Science and Technology, 47*(1), 15–26.

181. Salunkhe, D. K., (1974). *Storage, Processing and Nutritional Quality of Fruits and Vegetables.* CRC Press, Ohio, pp. 29–30.

182. Schimitt, H. A., & Weaver, C. M., (1982). Effects of laboratory scale processing on chromium and zinc in vegetables. *Journal of Food Science, 47*(5), 1693–1694.

183. Schroeder, H. A., (1971). Losses of vitamin and trace minerals resulting from processing and preservation of foods. *American Journal of Clinical Nutrition, 24*, 562–566.

184. Schwartz, S. J., & Von Elbo, J. H., (1983). High performance liquid chromatography of chlorophylls and their derivatives in freash and processed spinach. *Journal of Agricultural and Food Chemistry, 29*(3), 533–535.

185. Seet, S. T., & Brown, W. D., (1983). Nutritional quality of raw, precooked and canned albacore tuna (*Thunnus alalunga*). *Journal of Food Science, 48*, 288–289.

186. Severi, S., Bedogni, G., Manzieri, A. M., Poli, M., & Battistini, N., (1997). Effects of cooking and storage methods on the micronutrient content of foods. *European Journal of Cancer Prevention, 6*(Suppl. 1), 521–524.

187. Sharma, H. P., & Kumar, R. A., (2013). Health security in ethnic communities through nutraceutical leafy vegetables. *Journal of Environmental Research and Development, 7*(4), 1423–1429.

188. Shi, J., & Maguer, M. L., (2000). Lycopene in tomatoes: chemical and physical properties affected by food processing. *Critical Review in Food Science and Nutrition, 40*, 1–42.

189. Shitanda, D., & Wanjala, N. V., (2006). Effect of different drying methods on the quality of jute (*Corchorus olitorius* L.). *Drying Technology*, *24*(1), 95–98.

190. Sila, D. N., Van Buggenhout, S., Duvetter, T., Fraeye, I., De Roeck, A., & Van Loey, A., et al., (2009). Pectins in processed fruit and vegetables: Part II – structure function relationships. *Comprehensive Reviews in Food Science and Food Safety*, *8*, 86–104.

191. Singh, S., Raina, C. S., Bawa, A. S., & Saxena, D. C., (2006). Effect of pretreatments on drying and rehydration kinetics and color of sweet potato slices. *Drying Technology*, *24*, 1487–1494.

192. Sobukola, O. P., Dairo, O. U., Sanni, L. O., Odunewu, A. V., & Fafiolu, B. O., (2007). Thin layer drying process of some leafy vegetables under open sun. *Journal of Food Science and Technology International*, *2*, 13–35.

193. Somogyi, J. C., (1990). Influence of food preparation on nutritional quality: introductory remarks. *Journal of Nutritional Science and Vitaminology*, *36* (Suppl. l), SI-S6.

194. Somogyi, L. P., & Luh, B. S., (1986). Dehydration of fruits. In: *Woodroof, J. G. and Luh, B. S. Eds., Commercial Fruit Processing*. 2nd Ed., Westport: AVI Publishing Co., Inc, pp. 353–405.

195. Stahl, W., & Sies, H., (1996). Uptake of lycopene and its geometrical isomers is greater from heat-processed than from unprocessed tomato juice in humans. *Journal of Nutrition*, *122*, 2161–2166.

196. Stanley, D. W., Bourne, M. C., Stone, A. P., & Wismer, W. V., (1995). Low temperature blanching effects of chemistry, firmness and structure of canned green beans and carrots. *Journal of Food Science*, *60*, 327–333.

197. Steinmetz, K. A., & Potter. J. D., (1996). Vegetables, fruit, and cancer prevention: A review. *Journal of the American Dietetic Association*, *96*, 1027–1039.

198. Sugimura, T., (2002). Food and cancer. *Toxicology*, *181–182*, 17–21.

199. Swasdisevi, T.; Soponronnarit, S.; Prachayawarakorn, S.; & Phetdasada, W., (1999) Drying of chopped spring onion using fluidization technique. *Drying Technology*, *17*(6), 1191–1199.

200. Swatsitang, P., & Wonginyoo, R., (2008). Antioxidant Capacity of Vegetable Juices. *KKU Science Journal*, *36* (Suppl.), 83–94.

201. Tabassum, N., & Ahmad, F., (2011). Role of natural herbs in the treatment of hypertension. *Pharmacognosy Reviews*, *5*(9), 30–40.

202. Tabekhia, M. M., & Luh, B. H., (1980). Effect of germination, cooking and canning on phosphorus and phytate retention in dry bean. *Journal of Food Science*, *45*(2), 406–408.

203. Tamer, C. E., Suna, S., Incedayi, B., Sinir, G. Ö., & Copur, Ö. U., (2014). Impact of drying methods on physicochemical and sensory properties of apricot pestil. *Indian Journal of Traditional Knowledg*, *13*(1), 47–55.

204. Tanaka, M., & Kimura, S., (1988). Effect of heating conditions on protein quality of retort pouch meat. *Nippon Susain Gakkaishi*, *54*, 265–270.

205. Thane, C., & Reddy, S., (1997). Processing of fruits and vegetables: effect on carotenoids. *Nutrition & Food Science*, *2*, 58–65. 10.

206. Tijskens, L. M. M., Schijvens, E. P. H. M., & Biekman, E. S. A., (2001). Modelling the change in color of broccoli and green beans during blanching. *Innovative Food Science and Emerging Technologies*, *2*(4), 303–313.

207. Traub, D. A., (2002). The drying curve, Part 2:Applying the drying curve to your drying process. Accessed on 06 June 2016 URL: http://www.process heating.com./ext./resources./PH. /Home./Files./PDFs./1002ph, dryingfiles, pdf

208. Trewavas, A., & Stewart, D., (2003). Paradoxical effects of chemicals in the diet on health. *Current Opinion in Plant Biology, 6*,185–190.

209. Tseng, A., & Zhao, Y., (2012). Effect of Different Drying Methods and Storage Time on the Retention of Bioactive Compounds and Antibacterial Activity of Wine Grape Pomace (Pinot Noir and Merlot). *Journal of Food Science, 77*(9), H192–201.

210. Udofia, U. S., & Obizoba, I. C., (2005). Effects of sun and shade drying on nutrient and antinutrient content of some green leafy vegetables consumed in Uyo communities, Akwa Ibom State. *Journal of Biomedical Investigation, 3*(1), 1–5.

211. USDANDB (United States Department of Agriculture, Agricultural Research Service, National Nutrient Database). United States Department of Agriculture, Agricultural Research Service, National Nutrient Database for Standard Reference. Accessed on 02 June 2016, URL: https://ndb.nal.usda.gov/ndb/search

212. USEPA (US Environmental Protection Agency). Pesticides. Accessed on 02 June 2016, URL: https://www.epa.gov/pesticides

213. Vallejo, F., Tomás-Barberán, F. A., Garcia-Viguera, C., (2002). Glucosinolates and vitamin C content in edible parts of broccoli florets after domestic cooking. *European Food Research and Technology, 215*, 310–316.

214. Van Buran, J. P., Lee, C. Y., & Massey, L. M., (1982). Variation of vitamin concentration and retention in canned snap beans from three processing plants during two years. *Journal of Food Science, 47*, 1545–1548.

215. Van Buren, J. P., (1979). The chemistry of texture in fruits and vegetables. *Journal of Texture Studies, 10*(1); 1–23.

216. Veer, P., Jansen, M. C. J. F., Klerk, M., & Kok, F. J., (2000). Fruits and vegetables in the prevention of cancer and cardiovascular disease. *Public Health Nutrition, 3*, 103–107.

217. Verhoeven, D. T. H., Verhagen, H., Goldbohm, R. A., Brandt, P. A., & Poppel, G., (1997). A review of mechanisms underlying anticarcinogenicity by Brassica vegetables. *Chemico-Biological Interactions, 103*, 79–129.

218. Wagner, H. G., & Porter, A. E., (1973). The effect of maturity and variety on the content of the major organic acids of the green pea (*Pisum sativam*). *Journal of the Science of Food and Agriculture, 24*, (1), 69–75.

219. WHF (WORLDSHEALTHIESTFOODS). Broccoli. Accessed on 13 May 2016, URL: http://www.worldshealthiestfoods.org/genpage.php?tname=foodspice&dbid=9

220. Williams, D. C., Lim, M. H., Chen, A. O., Pangborn, R. M., & Whitaker, J. R., (1986). Blanching of vegetables for freezing:which indicator enzyme to choose. *Food Technology, 40*, 130–140.

221. Wu, A., & Chang, W. H., (1990). Influence of precooking on the firmness and pectic substances of three stem vegetables. *International Journal of Food Science and Technolog , 25*, 558–565.

222. Yamamoto, J. S., (1968). Stephenson, R. M. Free-flowing onion, garlic or horseradish. US Patent 3378380.

223. Yang, C. S. T., & Atallah, W. A., (1985). Effect of four drying methods on the quality of intermediate moisture lowbush blueberries. *Journal of Food Science, 50*(5), 1233–1237.

224. Youdim, K. A., & Joseph, J. A., (2001). A possible emerging role of phytochemicals in improving age-related neurological dysfunction: a multiplicity of effects. *Free Radical Biology & Medicine, 30*, 583–594.

225. Youdim, K. A., & Joseph, J. A., (2001). A possible emerging role of phytochemicals in improving age-related neurological dysfunctions: A multiplicity of effects. *Free Radical Biology & Medicine*, 30: 583–594.

226. Zhang, D., & Hamauzu, Y., (2004). Phenolics, ascorbic acid, carotenoids and antioxidant activity of broccoli and their changes during conventional and microwave cooking. *Food Chemistry, 88*, 503–509.

227. Zhong, T., & Lima, M., (2003). The effect of ohmic heating on vacuum drying rate of sweet potato tissue. *Bioresource Technology, 87*, 215–220.

228. Zia-ur-Rechman, Z., Islam, M., & Shah, W. H., (2003). Effect of microwave and conventional cooking on insoluble dietary fiber components of vegetables. *Food Chemistry, 80*, 237–240.

PART III

MANAGEMENT STRATEGIES IN AGRICULTURAL ENGINEERING

CHAPTER 7

ETHANOL PRODUCTION FROM DIFFERENT SUBSTRATES: EFFECTS ON ENVIRONMENTAL FACTORS AND POTENTIAL APPLICATIONS

LEELA WATI, ANNU GOEL, and KUSHAL RAJ

CONTENTS

7.1 INTRODUCTION

The word *alcohol* originally was referred to any fine power, however in 17th century, alchemists applied this term to the refined products of distillation. Ultimately, use of *alcohol* for ethanol was introduced in the mid of 18th century [38]. Ethanol is the systematic name defined by the International Union of Pure and Applied Chemistry (IUPAC) for a molecule with two carbon atoms (prefix "eth-"), having a single bond between them (suffix "-ane"), and an attached functional group- OH group (suffix "-ol"). During the mid-1800s, ethanol became one of the first structural formulas to be determined. Its molecular formula CH_3CH_2OH is abbreviated as C_2H_5OH. The name *ethanol* was coined as a result of a resolution that was adopted at the International Confer- ence on Chemical Nomenclature (ICCN) that was held in April 1892 in Geneva, Switzerland [2].

This chapter discusses effects of ethanol production from different substrates on environmental factors and potential applications.

7.2 PHYSICAL AND CHEMICAL PROPERTIES OF ETHANOL

Ethanol is a clear, colorless, flammable liquid with a pleasant smell. It has somewhat sweet flavor in dilute aqueous solution, but has a burning taste in more concentrated solutions. It boils at 78.5°C and has a density of 0.789 g/ml at 20°C. Its specific gravity is 0.79, which indicates that it is lighter than water. Since it is water-soluble, it is thoroughly miscible with water. It forms a constant-boiling mixture, or azeotrope, with water. Its vapor density is 1.59, which indicates that it is heavier than air. It burns with a smokeless blue flame that is not always visible under normal light. It has a wider flammable range than gasoline [13]. Due to Hydrogen bond- ing, it is highly hygroscopic (i.e., readily absorbs water from the air). It is a versatile solvent, miscible with water and with many organic solvents. The polar nature of the hydroxyl group causes ethanol to dissolve many ionic compounds, notably sodium and potassium hydroxides, magnesium chloride, calcium chloride, ammonium chloride, ammonium bromide and sodium bromide [33].

7.3 HISTORICAL BACKGROUND

Dried ethanol residues have been found on 9000 year old pottery in China, which indicates that Neolithic people (about 4000 BC) in this part of the world might have consumed alcoholic beverages [34]. The oldest kind of alcoholic drinks are wine and beer, remains of which have been found in tombs and settlements of early civilizations. There is a convincing evidence that the development of agriculture (regarded as the foundation of civilization) was based on the cultivation of grain for beer, as much as for bread. The first known brewery dates from about 3700 B.C. in Egypt and wine was discovered later in ancient Mesopotamia.

In 1796, Johann Tobias Lowitz obtained pure ethanol by filtering distilled ethanol through activated charcoal. It was only during 1800s that ethanol was used as fuel in engine [10]. The first major fuel-ethanol project was started in Brazil in 1975 by the name "*ProAlcool*" followed by similar programs in USA and Canada. The Government of India (GOI) on 1st January 2003 permitted blending of petrol with 5% ethanol and future target of 15% blending by 2017 and further to 20% by 2020. Ethanol contains 35% oxygen, which results in a complete combustion of fuel and lowers the emission of harmful gases. Ethanol produced from biomass resources has the potential to cut greenhouse gas emissions by 86%. Ethanol has a higher octane (ability to resist compression) rating than gasoline, enabling combustion engines to run at a higher compression ratio and thus giving a superior net performance. Emission of carbon monoxide and unburnt hydrocarbons that form smog can be reduced by the addition of ethanol to gasoline [24].

7.4 PRODUCTION PROCESS

Ethanol can be made synthetically from petroleum or by microbial conversion of biomass materials through fermentation and the latter is termed as bioethanol. Ethanol which is intended for industrial use is made by chemical synthesis, while ethanol intended for food use tends to be made by biological process, that is, fermentation. Fermentation of sugars into ethanol is one of the earliest organic reactions employed by humanity.

About 95% of the ethanol in the world is produced by fermentation and the rest by synthetic method.

7.4.1 CHEMICAL SYNTHESIS

Chemically, ethanol is synthesized either from acetaldehyde made from acetylene or by hydration of ethene [4]. A catalyst of phosphoric acid is used to ensure a fast reaction and ethanol is the only product.

$$C_2H_4 \quad + \quad H_2O \quad \rightarrow \quad C_2H_5OH$$
Ethene   Steam   Ethanol

The process is continuous, as long as ethene and steam are fed into one end of the reaction vessel, where ethanol is produced. These features make it an efficient process, but ethene is made from crude oil, which is a non-renewable resource and cannot be replaced once it is used up.

7.4.2 BIOLOGICAL PROCESS

In biological process, sugars from plant materials are converted into ethanol and carbon dioxide by fermentation. The enzymes found in microorganisms are the natural catalysts that can make this process happen. The production of ethanol or ethyl alcohol from starch or sugar-based feedstock is among man's earliest ventures into value-added processing [12]. Ethanol for use in alcoholic beverages, and the vast majority of ethanol for use as fuel, is produced by the fermentation process, by which microorganisms (yeast, fungi or bacteria) use monosaccharaides as a carbon source and convert to ethanol via glycolysis under anaerobic conditions. Unlike ethene, sugars from plant material are a renewable resource:

$$C_6H_{12}O_6 + 2ADP \rightarrow 2C_2H_5OH + 2CO_2 + 2\ ATP$$
Glucose   Ethanol

Theoretical yield from 1 g of glucose is 0.51 g of ethanol and 0.49 g of carbon dioxide. The ATP generated is used for cell processes. In ideal fermentation, about 95% of sugar is converted to ethanol and carbon dioxide, 1% is converted into cellular matter of the fermenting microbial cells, and 4% is converted into other products such as glycerol. The biological process of ethanol production is affected by microorganisms, substrates and environmental conditions.

7.5 ROLE OF MICROORGANISMS FOR ALCOHOL PRODUCTION

The production of alcohol involves active participation of microorganisms, most often yeasts while bacteria and fungi are also being used. Use of microbes accounts for about 10% of the ethanol production cost [48].

7.5.1 YEASTS

Yeasts are unicellular eukaryotic microorganisms that ferment variety of sugars from different sources into carbon dioxide and alcohol. The most commonly used yeast is *S. cerevisiae* because it can produce ethanol to a concentration as high as 18% in the fermentation broth [26]. Over 96% of the fermentation ethanol is produced by *S. cerevisiae* or species related to it [46]. *Saccharomyces* is also generally recognized as safe (GRAS) as a food additive for human consumption. The species or strains used in the fermentation play an important role in giving the final taste properties in alcoholic beverage production [17]. The metabolism of glucose in *S. cerevisiae* is by Embden-Meyerhof (EMP) pathway. Pyruvate formed during glycolysis is converted to acetaldehyde by pyruvate decarboxylate and acetaldehyde is converted into ethanol by acetaldehyde dehydrogenase. All the enzymes involved in fermentation are expressed constitutively and comprise 50% of cells' total protein. Pentose sugars are not fermented by *S. cerevisiae*.

The Pentose fermenting yeast strains *viz. Pichia stipitis, Candida shehatae* and *Pachysolan tannophilus* have the potential to metabolize both the hexose as well as pentose sugars and find application in cellulose based ethanol production. *Kluveromyces lactis* can be used to ferment whey into ethanol. Thermo-tolerant strains of non-conventional yeast like

Kluveromyces could be more suitable for ethanol production at industrial level because at high temperature process energy savings can be achieved through a reduction in cooling costs requiring no additional finance during downstream processing [37].

7.5.2 FUNGI AND BACTERIA

Other microorganisms capable of fermenting a great variety of sugars including hexoses and pentose are the fungi belonging to genera *Mucor, Rhizopus, Monilia, Neurospora and* Paecilomyces giving high ethanol yields when cultivated under anaerobic conditions. *Mucor indicus* can assimilate the inhibitors present in dilute-acid hydrolyzates [42]. The use of the fungus *Chalara parvispora* for ethanol production from pentose-containing materials was patented in 2006 by Holmgren and Sellstedt [20]. Fungal strains of *Aspergillus oryzae* are used in production to facilitate starch fermentation from rice.

Among bacteria, *Zymomonas mobilis* is an unusual Gram-negative microorganism that utilizes glucose anaerobically by Enter-Doudoroff (ED) pathway and ED pathway yields half as much ATP per mole of glucose in EMP pathway. As a consequence this, bacterium produces less biomass than yeast and yield 5–10% higher ethanol yield per glucose molecule fermented. *Z. mobilis* has been used in place of yeast in alcoholic fermentations [6]. It is also GRAS and is utilized for ethanol production from sugarcane juice in Brazil.

Fermenting bacteria have characteristic sugar fermentation patterns, *i.e.*, they can metabolize some sugars but not others. Xylose-fermenting thermophilic bacteria *Clostridium thermocellum* are prospective organisms to be co-cultured with cellulose hydrolyzing bacteria in order to directly convert pretreated lignocellulosic biomass into ethanol, process named consolidated bioprocessing (CBP). The fermentative index of the xylose-assimilating thermophilic bacteria *Thermophilus thermosaccharolyticum, T. ethanolicus* and *Bacillus stearothermophilus* has been reported by many researchers [28]. Engineering of *Escherichia coli* is another valuable bacterial resource for ethanol production. Genes encoding for pyruvate decarboxylate and acetaldehyde dehydrogenase have been cloned into

E. coli. This bacterium can ferment wide spectrum of sugars but ferment at narrow pH range and the cultures are less hardy as compared to yeast.

7.6 FEEDSTOCKS FOR FERMENTATIVE ETHANOL PRODUCTION

Several kinds of carbohydrate-rich raw materials are used for ethanol production. These can be classified into three groups of agricultural products: all sugar, starch and lignocellulose. The first raw material group, sugar refers to the fruit juices, sugarcane, sugar-beet and molasses. The second group includes starch from crops like cassava, cereals like rice, wheat, corn and tuber crops like potatoes and sweet potatoes. The third group, lignocellulose, covers waste materials from the harvesting of agricultural crops such as rice straw, corn cob and sugarcane waste. Ethanol for drinking purpose is mainly produced from sugar rich fruits and starchy substrates while ethanol as fuel is mainly produced from sugary by products like molasses, cellulosic and lignocellulosic substrates [25].

7.6.1 ETHANOL PRODUCTION FROM SUGAR RICH SUBSTRATES

Most fruits, berries like grapes, palm, banana, etc., and sugar crops like sugarbeet, sugarcane, sweet sorghum, etc., are good substrates for ethanol production. Fruit juices have balanced quantities of sugar, acid, tannin, nutritive salts as yeast nutrients and water to produce naturally stable, drinkable liquor. Sometimes amount of fermentable sugars is often low and need to be supplemented by a process called chaptalization in order to have sufficient alcohol levels in the finished product [35]. Sugarcane juice has about 12% fermentable sugars and can be used directly for fermentation. Brazil is second biggest ethanol producer in the world, where direct sugarcane juice is fermented in continuous process employing *Z. mobilis* [16]. Countries in Central America and Caribbean are suitable for sugarcane cultivation, and their ethanol production is increasing recently. Sugar beet mainly cultivated in European countries is processed to get sugar beet molasses that can also be fermented to ethanol [32].

Molasses from sugar industry, spent sulphite liquor, etc. can also be used for ethanol production. Molasses obtained after sugar recovery contains about 35% sucrose, 19% other reducing sugars and 14% other organic substances, thus the total fermentable sugars are nearly 50–55%. In India, molasses are most common substrate for ethanol production. Molasses is first suitably diluted, then treated with 0.5% by weight concentrated sulphuric acid at 70–95°C (mainly for the removal of calcium salts and acidification) and used for fermentation [40]. A solution of sucrose, to which yeast is added, is acted upon by an enzyme, invertase, which is present in yeast and acts as a catalyst to convert sucrose into glucose and fructose.

Invertase

$$C_{12}H_{22}O_{11} + H_2O \rightarrow C_6H_{12}O_6 + C_6H_{12}O_6$$

Sucrose   Glucose   Fructose

The glucose and fructose formed are then converted into ethanol and carbon dioxide by another enzyme, zymase, which is also present in yeast.

Zymase

$$C_6H_{12}O_6 \rightarrow 2C_2H_5OH + 2CO_2$$

Glucose   Fructose   Ethanol

Due to fluctuations in sugarcane productivity and hike in the price of sugarcane molasses, many of the Indian distilleries have switched over to alternate substrates for industrial alcohol production.

7.6.2 ETHANOL PRODUCTION FROM STARCHY SUBSTRATES

Ethanol can be fermented from many sources of starch, including corn, wheat, grain sorghum, barley and potatoes. Cereals like rice, wheat, corn, etc. contain 60–75% w/w starch which on hydrolysis produces glucose in the ratio of 9:10. Root and tuber crops like potato and sweet potato because of their high contents of starch also are potential feedstocks for ethanol production. Generally, starch is a mixture of α-amylose (20–30%; water-soluble linear polymer) and amylopectin (70–80% water insoluble branched polymer). Cereals typically consist of 70% amylopectin and 30% amylose. Amylose is essentially a linear polymer containing 70–2100 glucose units, 4 bonds and it has a double helical crystalline structure. The helix contains six D-glucose molecules per turn. Amylopectin in contrast to amylose is a highly branched structure with 4–6% of α-1,6 bonds at branch points. The average length of branch chain is 20–25 glucose units [41].

Starch is a high yield feedstock for ethanol production but to overcome food *vs* fuel debate in developing countries like in India only the underutilized portion of this can be diverted towards ethanol production. The bioconversion of starch into ethanol is not as simple as in case of sugary materials but its hydrolysis is required before fermentation [41]. Starch was traditionally hydrolyzed by acids, but hydrolysis using acid is not satisfactory because of undesirably colored and flavored breakdown products. Acid hydrolysis appears to be a totally random process which is not influenced by the presence of α-1,6-glucosidic linkages. Enzymatic hydrolysis, on the other hand, has the advantage of being highly specific, eco-friendly and formation of no fermentation inhibitory by-products [30, 39].

Among the various starch splitting enzymes, α-amylase and glucoamylase are widely used enzymes in the starch industry and are of commercial importance. Microorganisms like yeast, fungi, bacteria, actinomycetes and algae are able to produce α-amylase [47] but only the enzymes from fungal and bacterial sources have been employed for industrial applications. A number of companies are involved in supplying commercial grade enzymes for hydrolysis. Commercial enzymes used for the industrial hydrolysis of starch are produced mainly by *Bacillus amyloliquefaciens* and *B. licheniformis*. The fungal enzymes mainly produced by species of *Aspergillus* are preferred rather than those from bacteria as their action is easier to control [8].

Gelatinization, a predecessor to enzymatic hydrolysis has been widely used to speed up the starch hydrolysis. Most amylolytic enzymes are inactive on the native granules of starch. Therefore, starch must be thinned or gelatinized by heat treatment before it can be hydrolyzed into glucose. During gelatinization, the starch slurry is adjusted to approximately 15–30% dry solids and passed through a high-temperature heat exchanger (Jet cooker) that instantaneously raises the temperature of the slurry to the gelatinization temperature (~100°C). Under high temperature conditions, anhydrous starch granules begin to swell irreversibly and the amylose in the granules becomes soluble, resulting in a colloidal suspension or gel in aqueous phase due to breakage of surface-layer of swollen starch granules. This gel later becomes more susceptible to saccharifying agent.

The process has been considerably refined in recent years, leading to a very efficient process. United States mainly produces ethanol from corn, and this also makes the United States the world biggest ethanol producer [29]. In developing countries like India, many starch based ethanol plants have been set up. There are two production processes for starch ethanolic conversion: dry milling and wet milling. The main difference between the two is in the initial treatment of the grain.

7.6.2.1 Dry Milling

In dry milling, the entire starchy grain is first ground into flour, which is referred as "meal" and processed without separating out the various component parts of the grain. Reduction in particle size of starchy material is must to allow better penetration of enzymes. The meal is slurried with water to form a "mash" and processed at high-temperature (105°C using live steam) to gelatinize starch and to reduce bacteria levels ahead of fermentation. The partly gelatinized starch is passed into a series of holding tubes maintained at 100–105°C and held for 5 min to complete the gelatinization process. Gelatinized starch is readily liquefied by partial hydrolysis with enzymes and saccharified further by enzymatic hydrolysis. For optimum activity of enzymes calcium ions are added. The enzymatic activity, combined with the significant shear forces, bring about the hydrolysis of starch into dextrose, a simple sugar. Ammonia is added for pH control and as a nutrient to the yeast. Hydrolysis is completed in holding tanks at 90–100°C for 1 to 2 h. The hydrolyzed starch is transferred to fermenters where active yeast inoculum is added and the conversion of sugar to ethanol and carbon dioxide (CO_2) begins. The fermentation process generally takes about 40 to 50 hours. During this part of the process, the mash is agitated and kept cool to facilitate the activity of the yeast. After fermentation, the resulting fermented solution is transferred to distillation columns where ethanol is separated from the remaining "stillage".

7.6.2.2   Wet Milling

In wet milling, ethanol production requires only the starch portion of grain. The remaining protein, fat, fiber and other nutrients are returned to the global livestock and poultry feed markets. The grain is soaked or "steeped" in water and dilute sulfurous acid for 24 to 48 hours. This steeping facilitates the separation of the grain into its many component parts. After steeping, the grain slurry is processed through a series of grinders to separate the grain germ. The grain oil from the germ is either extracted on-site or sold to crushers and the remaining fiber, gluten and starch components are further segregated using centrifugal, screen and hydrocyclonic separators.

The steeping liquor is concentrated in an evaporator. This concentrated product, heavy steep water, is co-dried with the fiber component and is then sold as grain gluten feed to the livestock industry. Heavy steep water is also sold by itself as a feed ingredient and is used as a component in Ice Ban, an environmentally friendly alternative to salt for removing ice from roads. The gluten component (protein) is filtered and dried to produce the grain gluten meal co-product used as a feed ingredient in poultry boiler operations. The starch and any remaining water from the mash are fermented into ethanol. The fermentation process for ethanol is very similar to the dry mill process.

7.6.3   ETHANOL PRODUCTION FROM LIGNOCELLULOSIC MATERIALS

Lignocellulosic complex is the most abundant biopolymer on the Earth comprising about 50% of world biomass. Many lignocellulosic materials have been tested for ethanol production world over depending upon their availability. In general, prospective lignocellulosic materials for fuel ethanol production can be divided into six main groups [37]:

- *Crop residues*, for example, cane bagasse, corn stover, wheat straw, rice straw, rice hulls, barley straw, sweet sorghum bagasse, olive stones and pulp.
- *Hardwood, for example,* aspen, poplar.
- *Softwood*, for example, pine, spruce.

- *Cellulose wastes*, for example, newsprint, waste office paper, recycled paper sludge.
- *Herbaceous biomass*, for example, alfalfa hay, switchgrass, reed canary grass, coastal Bermuda grass, timothy grass.
- *Municipal solid wastes.*

Three main components—cellulose, hemicellulose, and lignin—make up the cell walls that form the tissue structure of lignocellulosic biomass [3]. The cellulose $(C_6H_{10}O_5)_n$ is a linear polymer of glucose, composed of thousands of molecules of anhydroglucose linked by β-(1, 4)-glycosidic bonds. The basic repeating unit is the disaccharide cellobiose. The term hemicellulose is a collective term used to represent a family of polysaccharides such as arabino-xylans, gluco-mannans, galactans, and others that are found in the plant cell wall and have different composition and structure depending on their source and the extraction method.

The most common type of polymers that belong to the hemicellulose family of polysaccharides is xylan. The molecule of a xylan involves 1–4 linkages of xylopyranosyl units with α-(4-O)-methyl-D-glucuronopyranosyl units attached to anhydroxylose units. The result is a branched polymer chain that is mainly composed of five carbon sugar monomers, xylose, and to a lesser extent six carbon sugar monomers such as glucose [27]. Lignin is the most complex natural polymer. It is an amorphous three-dimensional polymer with phenyl propane units as the predominant building blocks. More specifically p-coumarvl alcohol, coniferyl alcohol and sinapyl alcohol are the ones most commonly encountered. Lignin behaves as an insoluble three-dimensional network around cellulose and hemicellulose.

Converting lignocellulosic biomass to ethanol consists of three major unit processes: (i) pre-treatment involving the removal of lignin and disruption of crystalline structures of cellulose; (ii) saccharification for conversion of cellulose and hemicellulose into fermentable sugars; and (iii) fermentation of sugars into ethanol.

7.6.3.1 Pre-Treatment

Pre-treatment is an important step for the biochemical conversion of lignocellulosic materials into bioethanol. The purpose of the pre-treatment is to remove lignin, reduce cellulose crystallinity, and increase the porosity of the materials [27]. Pre-treatment must meet the following requirements:

- Improve the formation of sugars or the ability to subsequently form sugars by enzymatic hydrolysis.
- Avoid the degradation or loss of carbohydrate.
- Avoid the formation of by-products inhibitory to the subsequent hydrolysis and fermentation process.
- Be cost-effective.

Various physical, physicochemical, chemical, and biological processes have been used alone as well as in combination for pre-treatment of lignocellulosic materials [43].

7.6.3.1.1 Physical Pre-Treatment

Physical pre-treatment can increase the accessible surface area and size of pores, and decrease the crystallinity and degree of polymerization of cellulose. Different types of physical processes such as milling (e.g., ball milling, two-roll milling, hammer milling, colloid milling, and vibro energy milling) and irradiation (e.g., by gamma rays, electron beam or microwaves) can be used to improve the enzymatic hydrolysis or biodegradability of lignocellulosic waste materials. Milling can be employed to alter the inherent ultrastructure of lignocelluloses and degree of crystallinity, and consequently improve susceptibility to enzymatic hydrolysis [11].

7.6.3.1.2 Physicochemical Pre-Treatment

Pre-treatments that combine both physical and chemical processes are called physicochemical processes. The physicochemical pre-treatment step can be done by using acid, alkali, organic solvents, heat treatments, etc. and include steam explosion, ammonia fiber explosion, CO_2 explosion and Microwave treatment [49].

Steam explosion (Auto Hydrolysis)
Steam explosion is the most commonly used method for pre-treatment of lignocellulosic materials. In this method, biomass is treated with high-

pressure saturated steam and then the pressure is swiftly reduced, which makes the materials undergo an explosive decompression. Steam explosion is typically initiated at a temperature of 160–260°C for several seconds to a few minutes before the material is exposed to atmospheric pressure. The process causes hemicellulose degradation and lignin transformation due to high temperature, thus increasing the potential of cellulose hydrolysis.

The factors that affect steam explosion pre-treatment are residence time, temperature, particle size and moisture content. Optimal hemicellulose solubilization and hydrolysis can be achieved by either high temperature and short residence time (270°C, 1 min) or lower temperature and longer residence time (190°C, 10 min). Addition of H_2SO_4 (or SO_2) or CO_2 in steam explosion can effectively improve enzymatic hydrolysis, decrease the production of inhibitory compounds, and lead to more complete removal of hemicellulose.

The advantages of steam explosion pre-treatment include the low energy requirement compared to mechanical commination and no recycling or environmental costs. The conventional mechanical methods require 70% more energy than steam explosion to achieve the same size reduction. Steam explosion is recognized as one of the most cost effective pre-treatment processes for hardwoods and agricultural residues, but it is less effective for softwoods.

Limitations of steam explosion include destruction of a portion of the xylan fraction, incomplete disruption of the lignin–carbohydrate matrix, and generation of compounds that may be inhibitory to microorganisms used in downstream processes. Because of the formation of degradation products that are inhibitory to microbial growth, enzymatic hydrolysis and fermentation, pretreated biomass needs to be washed by water to remove the inhibitory materials along with water-soluble hemicelluloses. The water wash decreases the overall saccharification yields due to the removal of soluble sugars, such as those generated by hydrolysis of hemicellulose. Typically, 20–25% of the initial dry matter is removed by water wash.

Ammonia fiber explosion (AFEX)

Ammonia fiber explosion (AFEX) is another type of physicochemical pretreatment in which lignocellulosic materials are exposed to liquid ammonia at high temperature and pressure for a period of time, and then the pressure is swiftly reduced [44]. The concept of AFEX is similar to steam explosion. In a typical AFEX process, the dosage of liquid ammonia is 1–2 kg ammonia/kg dry biomass, temperature 90°C, and residence time 30 min. AFEX pre-treatment can significantly improve the saccharification rates of various herbaceous crops and grasses. It can be used for the pre-treatment of many lignocellulosic materials including alfalfa, wheat straw, wheat chaff, barley straw, corn stover, rice straw, municipal solid waste, softwood newspaper, kenaf newspaper, coastal Bermuda grass, switchgrass, aspen chips and bagasse. The AFEX pre-treatment does not significantly solubilize hemicellulose compared to acid pre-treatment and acid-catalyzed steam explosion. This treatment does not produce inhibitors for the downstream biological processes, so water wash is not necessary and does not require small particle size for efficacy.

To reduce the cost and protect the environment, ammonia must be recycled after the pre-treatment. In an ammonia recovery process, a super-heated ammonia vapor with a temperature up to 200°C can be used to vaporize and strip the residual ammonia in the pre-treated biomass and the evaporated ammonia may then be withdrawn from the system by a pressure controller for recovery.

CO_2 explosion

CO_2 explosion is similar to steam and ammonia explosion pre-treatment and is also used for pre-treatment of lignocellulosic materials. This treatment is more cost-effective than ammonia explosion and does not cause formation of inhibitory compounds that could occur in steam explosion.

Microwave pre-treatment

Microwave-based pre-treatment can be considered a physicochemical process since both thermal and non-thermal effects are often involved. Pre-treatments are carried out by immersing the biomass in dilute chemicals like acid/alkali/H_2O_2 and exposing the slurry to microwave radiation for residence time ranging from 5 to 20 min.

7.6.3.1.3   Chemical Pre-Treatment

Acid pre-treatment

The acid pre-treatment solubilizes the hemicelluloses fraction of biomass and makes cellulose more accessible to enzymes. This can be performed with concentrated or dilute mineral acids such as H_2SO_4 and HCl. Concentrated acids powerful agents for pre-treatment but are toxic, corrosive and hazardous and require reactors that are resistant to corrosion. In addition, the concentrated acid must be recovered after treatment to make the process economically feasible. Dilute acid treatment can be successfully developed for pre-treatment of lignocellulosic materials because this can achieve high reaction rates and significantly improve cellulose hydrolysis. The reaction occurs under less severe conditions and high xylan to xylose conversion yields are achieved which is necessary for overall process economics because xylan accounts for up to a third of the total carbohydrate in many lignocellulosic materials.

There are primarily two types of dilute acid pre-treatment processes: high temperature (>160°C), continuous-flow process for low solids loading (5–10% [weight of substrate/weight of reaction mixture]), and low temperature (<160°C), batch process for high solids loading (10–40%). Although dilute acid pre-treatment can significantly improve the cellulose hydrolysis, its cost is usually higher and neutralization of pH is necessary for the downstream enzymatic hydrolysis or fermentation processes.

Alkaline pre-treatment
Bases like sodium, calcium and ammonium hydroxides can also be used for pre-treatment of lignocellulosic materials and the effect of alkaline pre-treatment depends on the lignin content of the materials. The mechanism of alkaline treatment is the saponification of intermolecular ester bonds cross linking xylan hemicelluloses and other components, for example, lignin and other hemicellulose. The porosity of the lignocellulosic materials increases with the removal of the cross-links. Dilute NaOH treatment of lignocellulosic materials causes swelling, leading to an increase in internal surface area, a decrease in the degree of polymerization, a decrease in crystallinity, and separation of structural linkages between lignin and carbohydrates, and disruption of the lignin structure [14].

Ozonolysis
Ozone is a powerful oxidant that can be used to degrade lignin and hemicelluloses in many lignocellulosic materials such as wheat straw, bagasse, green hay, peanut, pine, cotton straw, and poplar sawdust. The degradation

is essentially limited to lignin and hemicellulose is slightly attacked, but cellulose is hardly affected. Ozonolysis pre-treatment has the following advantages: (1) it effectively removes lignin; (2) it does not produce toxic inhibitory compounds; (3) the reactions are carried out at room temperature and pressure. The drawback of this process is the requirement of large amount of ozone, making the process economically unviable.

Oxidative delignification

Lignin biodegradation can be catalyzed by the peroxidase enzyme with the presence of H_2O_2. The pre-treatment of cane bagasse with hydrogen peroxide greatly enhances its susceptibility to enzymatic hydrolysis. About 50% lignin and most hemicelluloses are solubilized by 2% H_2O_2 at 30°C within 8 h, and 95% efficiency of glucose production from cellulose can be achieved in the subsequent saccharification by cellulase at 45°C for 24 h.

Organosolv process

Organosolv process achieves high lignin removal with minimum cellulose loss. Numerous organic or aqueous organic solvent mixtures with inorganic acid catalysts (HCl or H_2SO_4) are used to break the internal lignin and hemicellulose bonds. The organic solvents used in the process include methanol, ethanol, acetone, ethylene glycol, triethylene glycol and tetrahydrofurfuryl alcohol. Organic acids such as oxalic, acetylsalicylic and salicylic acid can also be used as catalysts in the organosolv process. Solvents used in the process need to be drained from the reactor, evaporated, condensed and recycled to reduce the cost. Removal of solvents from the system is necessary because the solvents may be inhibitory to the growth of organisms, enzymatic hydrolysis, and fermentation.

7.6.3.1.4   Biological Pre-Treatment

In biological pre-treatment processes, microorganisms such as brown-, white- and soft-rot fungi are used to degrade lignin and hemicellulose in waste materials. Brown rots mainly attack cellulose, while white and soft rots attack both cellulose and lignin. White-rot fungi are the most effective basidiomycetes for biological pre-treatment of lignocellulosic materials.

Several white-rot fungi such as *Phanerochaete chrysosporium, Ceriporia lacerata, Cyathus stercolerus, Ceriporiopsis subvermispora, Pycnoporus cinnarbarinus* and *Pleurotus ostreaus* have been examined on different lignocellulosic biomass showing high delignification efficiency [21]. These white-rot fungi produce lignin-degrading enzymes, lignin peroxidases and manganese- dependent peroxidases, during secondary metabolism in response to carbon or nitrogen limitation. Both enzymes are present in the extracellular filtrates of many white-rot fungi for the degradation of wood cell walls. Other enzymes including polyphenol oxidases, laccases, H_2O_2 producing enzymes and quinone-reducing enzymes can also degrade lignin. The advantages of biological pre-treatment include low energy requirement and mild environmental conditions. However, the rate of hydrolysis in most biological pre-treatment processes is very low.

7.6.3.2   Hydrolysis/Saccharification

It is the conversion of cellulose and hemicellulose to fermentable sugars, which can be carried out either by chemical or biological methods. Chemical hydrolysis can be carried out though use of mineral acids while biological methods include use of enzymes, i.e., cellulases and hemicellulases for hydrolysis of cellulose and hemicelluloses.

7.6.3.2.1   Acid Hydrolysis

Acid hydrolysis is only applicable when feedstock has been pretreated according to the dilute acid process. Both dilute and concentrated acid options are available for hydrolyzing pretreated feedstock. At this stage, higher temperature (about 215°C) and dilute acid (4%) are used for converting cellulose to glucose. The concentrated acid process has a very high sugar yield (90%), can handle diverse feedstock, is relatively rapid (10–12 hours in total), and causes small degradation. However, the equipment required is more expensive when compared to dilute acid hydrolysis [18].

7.6.3.2.2 Enzymatic Hydrolysis

Enzymatic hydrolysis of cellulose is carried out by cellulases which are highly specific enzymes. The products of the hydrolysis are usually reducing sugars including glucose [45]. Utility cost of enzymatic hydrolysis is low compared to acid or alkaline hydrolysis because enzyme hydrolysis is usually conducted at mild conditions (pH 4.8 and temperature 45–50°C) and does not have a corrosion problem. Both bacteria and fungi can produce cellulases for the hydrolysis of lignocellulosic materials. These microorganisms can be aerobic or anaerobic, mesophilic or thermophilic. Bacteria belonging to *Clostridium, Cellulomonas, Bacillus, Thermomonospora, Ruminococcus, Bacteriodes, Erwinia, Acetovibrio, Microbispora,* and *Streptomyces* can produce *cellulases.* Many cellulolytic bacteria, particularly *Clostridium thermocellum* and *Bacteroides cellulosolvens* produce cellulases with high specific activity but they do not produce high enzyme titres. Moreover the anaerobes have a very low growth rate and require anaerobic growth conditions; therefore cellulase production has been focused mostly on fungi belonging to species of *Trichoderma, Aspergillus* and *Penicillium.*

Cellulases are usually a mixture of several enzymes. Three major groups of cellulases are involved in the hydrolysis process: (a) endoglucanase (EG, endo-1,4-D-glucanohydrolase, or EC 3.2.1.4) which attacks regions of low crystallinity in the cellulose fiber, creating free chain-ends; (b) exoglucanase or cellobiohydrolase (CBH, 1,4-β-D-glucan cellobiohydrolase, or EC 3.2.1.91.) which degrades the molecule further by removing cellobiose units from the free chain-ends; (c) β-glucosidase (EC 3.2.1.21) which hydrolyzes cellobiose to produce glucose [9, 44]. Most commercial glucanases are produced by *Trichoderma reesei* and β-D-glucosidase is produced from *Aspergillus niger*. In addition to the three major groups of cellulase enzymes, there are also a number of ancillary enzymes that attack hemicellulose, such as glucuronidase, acetylesterase, xylanase, β-xylosidase, galactomannanase and glucomannanase. Many fungal species such as *Trichoderma, Aspergillus* and *Penicililum* are reported to secrete large amount of extracellular hemicellulases. During the enzymatic hydrolysis, cellulose is degraded by the cellulases to reducing sugars that can be fermented by yeasts or bacteria to ethanol.

a. Improving enzymatic hydrolysis

The factors that affect the enzymatic hydrolysis of cellulose include substrates, cellulase activity, and reaction conditions (temperature, pH, as well as other parameters). Improvement in the yield and rate of the enzymatic hydrolysis can be made by optimizing the hydrolysis process and enhancing the cellulase activity.

b. Substrates

Substrate concentration is one of the main factors that affect the yield and initial rate of enzymatic hydrolysis of cellulose. At low substrate levels, an increase of substrate concentration normally results in an increase of the yield and reaction rate of the hydrolysis. However, high substrate concentration can cause substrate inhibition, which substantially lowers the rate of the hydrolysis, and the extent of substrate inhibition depends on the ratio of total substrate to total enzyme. The susceptibility of cellulosic substrates to cellulases depends on the structural features of the substrate including cellulose crystallinity, degree of cellulose polymerization, surface area, and content of lignin. Lignin interferes with hydrolysis by blocking access of cellulases to cellulose and by irreversibly binding hydrolytic enzymes. Therefore, removal of lignin can dramatically increase the hydrolysis rate.

c. Cellulase

Increasing the dosage of cellulases in the process, can enhance the yield and rate of the hydrolysis to a certain extent, but would significantly increase the cost of the process. Enzymatic hydrolysis of cellulose consists of three steps: adsorption of cellulase enzymes onto the surface of the cellulose, the biodegradation of cellulose to fermentable sugars, and desorption of cellulase. Cellulase activity decreases during the hydrolysis. The irreversible adsorption of cellulase on cellulose is partially responsible for this deactivation. Addition of surfactants during hydrolysis is capable of modifying the cellulose surface property and minimizing the irreversible binding of cellulase on cellulose. The surfactants used in the enzymatic hydrolysis include nonionic Tween 20, 80, polyoxyethylene glycol, Tween 81, Emulgen 147, amphoteric Anhitole 20BS, cationic Q-86W, sophorolipid, rhamnolipid, and bacitracin. Inhibitory effects can be there with cationic Q-86W at high concentration and anionic surfactant Neopelex

F-25. Nonionic surfactants are therefore believed to be more suitable for enhancing the cellulose hydrolysis.

d. End-product inhibition of cellulase activity
Cellulase activity is inhibited by cellobiose and to a lesser extent by glucose. Several methods are there to reduce the inhibition, including the use of high concentrations of enzymes, the supplementation of β-glucosidases during hydrolysis, and the removal of sugars during hydrolysis by ultrafiltration or simultaneous saccharification and fermentation (SSF). The SSF process reduces the inhibition of end products of hydrolysis. In the process, reducing sugars produced in cellulose hydrolysis or saccharification is simultaneously fermented to ethanol, which greatly reduces the product inhibition to the hydrolysis.

The high cost of enzymes is currently the greatest challenge in this processing step. Although current world-leading enzyme suppliers have reduced the price of enzymes about 20- to 30-fold, the cost for enzymes is still the most expensive part of the entire bioethanol process. Hyper cellulase producing strains of microorganisms have been developed. An important approach to reduce the cost for enzyme hydrolysis is to develop an efficient pre-treatment method to reduce the enzyme dosage and enhance the yield of simple sugars. Sugar yield is typically less than 20% without pre-treatment, whereas yield after pre-treatment often exceeds 90% [7].

7.6.3.3 Fermentation

During fermentation, both pentose and hexose sugars must be converted into ethanol under optimum conditions. The conventional organisms, *S. cerevisiae* and *Z. mobilis* ferment only hexose sugars, that is, glucose therefore co-cultures of *S. cerevisiae* and pentose fermenting yeasts *viz. Pichia stipitis* or *Pachysolen tannophilus* are used. Engineered microbes like *E. coli* have also been developed which can ferment both pentose and hexose sugars. In order to keep distillation costs low, the appropriate microorganism is selected based on the need to achieve high ethanol yield while also withstanding inhibition from accumulating toxic substances and autointoxication from increasing ethanol concentration. Based on the different combinations of technologies adopted at the pre-treatment, hydrolysis, and fermentation stages of ethanol synthesis, several integrated technologies have been developed.

7.6.3.4   Separate Hydrolysis and Fermentation (SHF)

The bioconversion of cellulosic biomass into ethanol can be a sequential process where the hydrolysis of cellulose and the fermentation are carried out in different units. This configuration is known as separate hydrolysis and fermentation (SHF). The separation of hydrolysis and fermentation offers various processing advantages and opportunities. The advantage of SHF is the ability to carry out each step under optimal conditions, that is, enzymatic hydrolysis at 45–50°C and fermentation at about 30°C. It is also possible to run fermentation in continuous mode with cell recycling. The major drawback of SHF is that the sugars released inhibit the enzymes during hydrolysis [15].

7.6.3.5   Simultaneous Saccharification and Fermentation (SSF)

In an effort to reduce the production cost of cellulose based ethanol, saccharification and fermentation are carried out in a single unit. This variant is known as simultaneous saccharification and fermentation (SSF). This has further been improved to include the co-fermentation of multiple sugar substrates, that is, simultaneous saccharification and co-fermentation (SSCF) [50]. In SSF and SSCF, the glucose produced is immediately consumed by the fermenting microorganism, which avoids end-product inhibition of β-glucosidase. The ethanol produced can also act as an inhibitor in hydrolysis but not as strongly as cellobiose or glucose. Another advantage of SSF and SSCF compared with SHF is the process integration obtained when hydrolysis and fermentation are performed in one reactor; the numbers of reactors needed are reduced. The temperature of about 35°C in SSF is a compromise, but the development of thermo tolerant yeast strains capable of fermenting upto 45°C can improve the performance of SSF. A major disadvantage of SSF is the difficulty in recycling and reusing the yeast since it will be mixed with lignin residue.

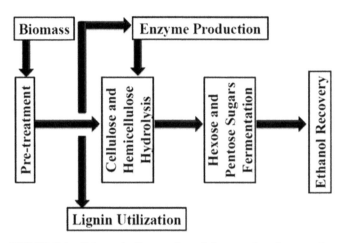

FIGURE 7.1   Schematic diagram for cellulose to ethanol conversion.

The cellulose based ethanol production (Figure 7.1) is not yet commercialized due to various technological barriers though worldwide some cellulose based plants have been tested. There is a need to develop a process that can bring economic advantage not only for fuel and power industry but also benefit the environmental rehabilitation and balance issues and cause. The enzymes can be produced in situ and lignin recovered after pre-treatment can be used for various industries.

7.7   EFFECTS OF ENVIRONMENTAL FACTORS ON ETHANOL PRODUCTION

The sugar solution (mash) obtained from sugary/starchy/cellulosic biomass is heavily inoculated with suitable organism pre grown in suitable media under sterilized conditions. A fast growing strain of *S. cerevisiae* is commonly used. The inoculum is prepared under sterilized conditions in two or more stages and inoculated at the level of 1×10^6 cells/ml alternatively commercially available yeast cakes can be used. Some of the companies are supplying compressed yeast preparations for ethanolic fermentations. Nitrogen source in the form of urea or ammonium sulphate and suitable phosphorous source is added as yeast nutrient for carrying out ethanol fermentation.

Several factors right from the composition of substrate to the type of organism, pH, temperature, ethanol concentration and nutrients are known to affect the production process. Yeasts are facultative organisms, while ethanol production is largely anaerobic some oxygen is needed by the yeast to synthesize some sterols and unsaturated fatty acid membrane components. Initially aeration is provided to promote yeast growth but anaerobiosis is soon established to promote fermentation and prevent oxidation of ethanol into CO_2 and H_2O. Many strains of *S. cerevisiae* can attain ethanol concentration of 12–14% but the bacterium *Z. mobilis* is less ethanol tolerant as compared to *S. cerevisiae*. Interest has been to produce ethanol tolerant yeast strains with a view to improve productivity. Yeast enzymes are sensitive to ethanol and plasma membrane phospholipid composition is important for ethanol tolerance Alcohol tolerance can be increased by adding unsaturated fatty acids in the medium. Yeast can grow and ferment preferably at pH value of 3–6. High pH favors glycerol formation and low pH disfavors the contaminating bacteria. Temperature optimum is different for different strains of ethanologenic organisms and use of thermo tolerant organisms is preferred over mesophilic organisms to save cooling cost. Batch fermentation is the common process for ethanol fermentation that takes about 48 h under control conditions until yeast stops growing due to high ethanol concentration [1]. Normally high concentration of ethanol keeps away the contaminants but acetic acid and lactic acid bacteria are the common contaminants during ethanol fermentations.

7.8 DOWNSTREAM PROCESSING

Like any fermentation reaction, the ethanol fermentation is not 100% selective because feedstocks used are of diverse composition. Many side products such acetic acid, glycols, esters, organic acids, higher alcohols and many other products are formed to a considerable extent and need to be removed during the purification of the ethanol. The fermentation takes place in aqueous solution and the resulting solution after fermentation has an ethanol content of 8–15%. The ethanol is subsequently recovered and purified by a combination of distillation and adsorption techniques. The recovery of ethanol is an energy intensive process.

7.8.1 DISTILLATION

Distillation is the most dominant and recognized industrial purification technique of ethanol. It relies on differences in the boiling points of the component liquids to be separated. The basic principle is that by heating a mixture, low boiling point components are concentrated in the vapor phase. By condensing this vapor, more concentrated less volatile compound is obtained in liquid phase. In ethanol production, a distillation tower is designed to separate water and ethanol effectively. Water is obtained from the bottom of the tower and ethanol is obtained from the top of the tower. Fractional distillation can concentrate ethanol to 95.6% by volume (89.5 mole %). This mixture is an azeotrope with a boiling point of 78.1°C (172.6°F), and cannot be further purified by distillation. Addition of an entraining agent, such as benzene, cyclohexane, or heptane, allows a new ternary azeotrope comprising the ethanol, water, and the entraining agent to be formed. This lower-boiling ternary azeotrope is removed preferentially, leading to water-free ethanol. Apart from distillation, ethanol may be dried by addition of a desiccant, such as molecular sieves, cellulose, and cornmeal. The desiccants can be dried and reused.

7.8.2 MOLECULAR SIEVES AND DESICCANTS

Molecular sieves can be used to selectively absorb water from the 95.6% ethanol solution. Synthetic zeolite in pellet form, as well as a variety of plant-derived absorbents, including cornmeal, straw and sawdust can be used. The zeolite bed can be regenerated essentially an unlimited number of times by drying it with a blast of hot carbon dioxide. Cornmeal and other plant-derived absorbents cannot readily be regenerated, but where ethanol is made from grain, they are often available at low cost. Absolute ethanol produced this way has no residual benzene, and can be used for various purposes [23].

7.8.3 MEMBRANES AND REVERSE OSMOSIS

Membranes are often used in hybrid membrane distillation process. This process uses a pre-concentration distillation column as first separating step. The further separation is then accomplished with a membrane operated either in vapor permeation or pervaporation mode. Vapor permeation uses a vapor membrane feed and pervaporation uses a liquid membrane feed. Membranes -based separations are not subject to the limitations faced as water-ethanol azeotrope because separation is not based on vapor-liquid equilibrium

7.9   WASTE DISPOSAL

Wastewater effluent after recovery of ethanol cannot be discharged directly on land or water body due to high BOD and COD. It is treated anaerobically to recover methane gas and the effluent left after biomethanation is discharged after suitable pre-treatment. The yeast biomass from sugarcane and molasses based fermentations is recycled while from starch based fermentations the stillage is sent through a centrifuge that separates the coarse grain from the solubles. The solubles are then concentrated to about 30% solids by evaporation, resulting in Condensed Distillers Solubles (CDS) or "syrup." The coarse grain and the syrup are then dried together to produce dried distillers grains with solubles (DDGS), a high quality and nutritious livestock feed. The CO_2 released during ethanol fermentation is captured and sold for use in carbonating soft drinks and beverages and the manufacture of dry ice.

7.10   ETHANOL ANALYSIS AND QUALITY ASSURANCE

7.10.1   GAS CHROMATOGRAPHY

Gas chromatography (GC) is an analytical technique for the volatile and semi-volatile compounds. Ethanol analysis can be done with GC since impurities in ethanol are basically volatile as well as ethanol itself [5, 19, 36]. A sample is vaporized at an injection port by heat. The sample vapor is sent to column packed with adsorbent or absorbent. There are many kinds of coatings for column. A coating should be chosen depending on the target compounds. Inside column, each component in sample is separated depending on its physical and chemical property. At the end of column the concentration of each compound is measured by a detector. There are many kinds of detectors. Each detector has advantages and disadvantages. Thus, a detector should also be chosen carefully to detect target compounds. Gas chromatography-mass spectrometry (GC-MS) is an integrated system of two analytical equipments. Gas chromatography separates analytes and mass spectrometry identifies them. GC-MS accelerates ethanol analysis with its simultaneous separation and identification capacities.

7.10.2 HIGH PERFORMANCE LIQUID CHROMATOGRAPHY

High performance liquid chromatography (HPLC) is an analytical technique which utilizes liquid as the mobile phase instead of gas of GC. Samples are not heated at the injection port. Thus, non-volatile compounds or heat sensitive compounds can be analyzed with HPLC. While HPLC is more comprehensive than GC in terms of sample limitation, it is still expensive and less sensitive comparing to GC [29].

7.10.3 INFRARED SPECTROSCOPY

Infrared spectroscopy (IR) is an analytical technique utilizing infrared adsorption. Infrared radiations with different wavelengths are passed through a liquid sample. Infrared is adsorbed by a compound, and the absorbability of infrared varies among different compounds and different infrared wavelengths. Samples are identified by comparing absorbability of infrared. IR does not have as high resolution as GC or HPLC. However, the equipment is relatively cheap and analysis is simple and quick. Thus, it utilizes more for quality assurance [22] and classification purposes [31].

7.10.4 OLFACTOMETRY

Olfactometry is a sensory analysis usually coupled with GC. For a typical GC-Olfactometry (GCO) system, a GC column is connected to a separator where analytes are separated to two ways, olfactometry and a detector such as FID, PID, and MS. Olfactometry is a simple system which is just an open-end column, and a panelist sniffs analytes coming from the column. The panellist records the odor character and intensity of the analyte which correspond with a peak in chromatogram. Olfactometry provides flavors data rather than stoichiometric chemical data. It is utilized for alcoholic beverage analysis to develop its flavor.

7.11   TYPES OF ETHANOL

Ethanol is produced in a range of purities for various applications and concentration of ethanol as adjusted accordingly. There are three types of ethanol as follows:

- *Absolute Alcohol:* Absolute or anhydrous alcohol is the ethanol with lowest water content ranging from 1% to a few parts per million (ppm) levels. Absolute alcohol is not intended for human consumption. It is used as a solvent for laboratory and industrial applications and as fuel alcohol. Spectroscopic ethanol is an absolute ethanol with a low absorbance in ultraviolet and visible light which is fit for use as a solvent in ultraviolet-visible spectroscopy. Pure ethanol is classed as 200 proof in the U.S. system and is equivalent to 175 proof in the UK system.
- *Rectified Spirit:* Rectified spirit is an azeotropic of 96% ethanol containing 4% water. It is used in laboratories instead of anhydrous ethanol for various purposes. Wine spirit contains about 94% ethanol (188 proof) and the impurities are different from laboratory spirit.
- *Denatured Alcohol:* Most industrial alcohol is denatured to prevent its use as beverage. Denatured alcohol contains 1–2% of poisonous substances such as denatonium benzoate and toxins such as methanol, naphtha, and pyridine that render it unfit to drink.

7.12   GRADES OF ALCOHOL

Several grades of ethanol are available in marketplace today. They differ primarily in the amount of impurities present. Generally, as we move down the chain to a lower quality ethanol, higher amount of impurities are present.

- **The Purest**: The highest level of ethanol purity is "GNS" or grain neutral spirits of beverage quality. This is produced only from fermentation grain sources, typically corn, wheat or broken rice. Based on the type of grain processed and the enzymes used to convert starch into sugar, different quantities and types of impurities are produced. Since GNS is intended for human consumption, it must pass stringent organoleptic analysis to meet all quality requirements.
- **Purer**: "FCC Grade" ethanol (Food Chemical Codex) is used in food applications. This can be either of fermentation or synthetic origin.
- **Pure**: "Industrial Grade" ethanol can be of either biological or synthetic origin. It has most of the impurities removed.
- **Not-as-pure**: The highest level of impurities is found in "fuel grade" ethanol that is produced via the fermentation process. It is added to gasoline as an octane enhancer/extender and to reduce carbon monoxide and nitrous oxide pollution. Since it is of biological origin, the impurities burn easily in a combustion engine.

7.13   CONCLUSIONS

Ethanol production has a long history because it is a known feedstock for countless chemicals and industrial purposes besides its use for potable purpose. Ethanol is used extensively as a solvent in the manufacture of varnishes and perfumes, as a preservative for biological specimens, in the preparation of essences and flavorings, in many medicines and drugs, as a disinfectant and as a fuel and gasoline additive. Ethanol burns more cleanly than gasoline fuels therefore fuel ethanol has become a very important agricultural product over the past two decades. Technological innovations are needed to produce this chemical from cheaper easily available renewable substrates to meet increasing demand of this chemical as biofuel.

7.14   SUMMARY

Ethanol is an alcohol produced through the action of microorganisms on sugars. It is the same alcohol produced by fermentation reaction of yeast in alcoholic beverages. Fuel alcohol is the ethanol that has been concentrated to remove water and blended with other components to render it undrinkable for use as biofuel. Fermentative ethanol production requires sugar solution prepared from various raw materials, efficient microorganisms and suitable environment. The final product can be recovered from fermented mash after suitable processing and can find various applications as beverage, in chemical industry, pharmaceutical applications and as biofuel.

KEYWORDS

- Absolute alcohol
- Acetaldehyde
- Acetaldehyde dehydrogenase
- *Acetovibrio*
- Acid hydrolysis
- Aerobic
- Aerobic microorganisms
- Agricultural crops
- Agricultural products
- Agriculture
- Alcohol
- Alcohol production
- Alcohol tolerance
- Alcoholic beverages
- Ammonia fiber explosion
- Anaerobic
- Anaerobic conditions
- Anaerobic microorganisms
- Ancillary enzymes
- Anhydrous alcohol
- Anhydrous starch
- Anhydroxylose units
- *Aspergillus*
- *Aspergillus niger*
- *Aspergillus oryzae*
- Azeotrope
- *Bacillus*
- *Bacillus amyloliquefaciens*
- *Bacillus licheniformis*
- *Bacillus stearothermophilus*
- Bacteria
- *Bacteriodes*
- Bacterium
- *Bacteroides cellulosolvens*
- Bbiological pre-treatment
- Bioconversion
- Bioethanol
- Biological
- Biological process
- Biomass
- Biomass materials
- Carbohydrate
- Carbon dioxide
- Carbon source
- Cellobiohydrolase
- Cellulase activity
- Cellulolytic bacteria
- *Cellulomonas*
- Cellulose
- Cellulose waste
- Cereals
- *Ceriporia lacerata*
- *Ceriporiopsis subvermispora*
- *Chalara parvispora*
- Chaptalization

REFERENCES

1. Achrekar, J., (2012). *Fermentation Biotechnology*. Wisdom Press New Delhi, pp. 210.
2. Armstrong, H., (1892). The International Conference on Chemical Nomenclature. *Nature, 46*, 56–59.
3. Badger, P. C., (2002). Ethanol from Cellulose: A General Review. In: *Janick, J. and Whipkey, A. Eds., Trends in new crops and new uses, Proceedings of the fifth National Symposium New Crops and New Uses: Strength in Diversity*. ASHS Press, Alexandria, VA, pp. 17–21.
4. Berthelot, M., (1855). *Sur la formation de l'alcool au moyen du bicarbured'hydrogène* (On the formation of alcohol by means of ethylene). *Annales de chimieet de physique, 43*, 385–405.
5. Campo, E., Cacho, J., & Ferreira, V., (2007). Solid phase extraction, multidimensional gas chromatography mass spectrometry determination of four novel aroma powerful ethyl esters: Assessment of their occurrence and importance in wine and other alcoholic beverages. *Journal of Chromatography, 1140*, 180–188.
6. Cardona, C. A., & Sanchez, O. J., (2007). Fuel ethanol production: process design trends and integration opportunities. *Bioresource Technology, 98,* 2415–2457.
7. Chandra, P. R., Bura, R., Mabee, W. E., Berlin, A., Pan, X., & Saddler, J. N., (2007). Substrate pretreatment: the key to effective enzymatic hydrolysis of lignocellulosics. *Advances in Biochemical Engineering/Biotechnology, 108*, 67–93.
8. Chaplin, M., (2014). The use of enzymes in starch hydrolysis: *In:Enzyme Technology*, Cambridge University Press, pp. 147.
9. Coughlan, M. P., & Ljungdahl, L. G., (1988). Comparative biochemistry of fungal and bacterial cellulolytic enzyme system. In. Aubert, J. P. and Beguin, P. M. (Eds) *Biochemistry and Genetics of Cellulose Degradation*, New York: Academic Press, pp. 11–30.
10. Couper, A. S., (1858). On a new chemical theory. *Philosophical magazine, 16*, 104–116.
11. da Silva, A. S., Inoue, H., Endo, T., Yano, S., & Bon, E. P., (2010). Milling pretreatment of sugarcane bagasse and straw for enzymatic hydrolysis and ethanol fermentation. *Bioresource Technology, 101*(19), 7402–7409.
12. Dudley R., (2004). Ethanol, fruit ripening and the historical origin of alcoholism in primate frugivory *Integrative and Comparative Biology, 44*, 315–323.
13. Farrell, A. E., Plevin, R. J., Turner, B. T., Jones, A. D., O'Hare, M., & Kammen, D. M., (2006). Ethanol Can Contribute to Energy and Environmental Goals. *Science, 311*, 506–508.
14. Goel, A., & Wati, L , (2013). Ethanol Production from Rice (*Oryza sativa*) straw biomass by separate hydrolysis and fermentation *Pure and Applied Microbiology, 7*, 3213–3218.
15. Goel, A., & Wati, L., (2013). Ethanol production from lignocellulosic materials. *International Journal of Innovations in Bio-Sciences, 13*, 111–114.

16. Goldemberg, J., Coelgo, S. T., & Guardabassi P., (2008). The sustainability of ethanol next term production from previous term sugarcane. *Energy Policy, 36*(6), 2086–2097.

17. Gray, K. A., Zhao, L., & Emptage., M., (2006). Bioethanol. *Current Opinion in Chemical Biology, 10*, 141–146.

18. Hamelinck, C. N., Hooijdonk, G. V., & Faaji, A. P. C., (2005). Ethanol from lignocellulosic biomass: techno-economic performance in short-, middle- and long-term. *Biomass and Bioenergy, 28*, 384–410.

19. Hida, Y., Kudo, K., Nishida N., & Ikeda, N., (2001). Identification of reddish alcoholic beverages by GC/MS using aroma components as indicators. *Legal Medicine, 2*(**4**), 237–240.

20. Holmgren, M., & Sellstedt, A., (2006). Producing ethanol through fermentation of organic starting materials, involves using fungus *e.g.*, *Chalaraparvispora*, capable of metabolizing pentose compounds. Patent SE527184, pp. 51.

21. Kumar, P., Barrett, D. M., Delwiche, M. J., & Stroeve, P., (2009). Methods for pretreatment of lignocellulosic biomass for efficient hydrolysis and biofuel production. *Industrial and Engineering Chemistry Research, 48*, 3713–3729.

22. Lachenmeier, D. W., (2007). Rapid quality control of spirit drinks and beer using multivariate data analysis of Fourier transform infrared spectra. *Food Chemistry, 101*(2), 825–832.

23. Lodgsdon, J. E., (1991). Ethanol. In: *Howe-Grant, M., Kirk, R. E., Othmer, D. F. and Kroschwitz, J. I. (Eds). Encyclopedia of Chemical Technology 9*. 4ᵗʰ Ed., New York: Wiley, pp. 817.

24. Mills, G. A., & Ecklund, E. E., (1987). Alcohols as Components of Transportation Fuels. *Annual Review of Energy, 12*, 47–80.

25. Mogg, R., (2004). Biofuels in Asia: Thailand relaunches gasohol for automotive use. *Refocus, 5*(3)**,** 44–47.

26. Morais, P. B., Rosa, C. A., Linardi, V. R., Carazza, F., & Nonato, E., (1996). Production of fuel alcohol by *Saccharomyces* strains from tropical habitats. *Biotechnology Letters, 18*, 1351–56.

27. Mosier, N. C., Dale, B., Elander, R., & Lee, Y. Y., (2005). Holtapple, M.;Ladisch, M. R. Features of technologies for pretreatment. *Bioresource Technology, 9*, 673–686.

28. Ogier, J. C., Ballerini, D., Leygue,J. P., Rigal, L., & Pourquie, J., (1999). Ethanol production from lignocellulosic biomass. *Oil and Gas Science and Technology/Revuedel Institute Francaisdu Petrol, 54* (1), 67–94.

29. Onuki, S., Koziel, J. A., Leeuwen, J. V., Jenks, W. S., Grewell, D., & Cai, L., (2015). Ethanol Production, Purification and analysis techniques: a review.*In: Agricultural and Biosystems Engineering Conference Proceedings and Presentations*. Paper 68, 2009,Accessed on 26 November, URL: http://lib.dr.iastate.edu/abe_eng_conf/68

30. Pandey, A., Nigam, P., Soccol, C. R., Soccol, V. T., Singh, D., & Mohan, R., (2000). Advances in microbial amylases. *Biotechnology and Applied Biochemistry, 31*, 135–152.

31. Pontes, M. J. C., Santos, S. R. B., Araújo, M. C. U., Almeida, L. F., Lima, R. A. C., & Gaião, E. N., et al., (2006). Classification of distilled alcoholic beverages and verification of adulteration by near infrared spectrometry. *Food Research International, 39* (2), 182–189.

32. Power, N., Murphy, J. D., & McKeogh, E., (2008). What crop rotation will provide optimal first generation ethanol production in Ireland, from technical and economical perspective? *Energy, 33*, 385–399.

33. Rahway, N. J., Windholz, M., & Budavari, S., (1976). Stroumtsos, L.Y. and Fertig, M.N.Eds.,*The Merck index: an encyclopedia of chemicals and drugs.* 9th Ed., U.S.A: Merck, pp. 678.

34. Roach, J., 9,000-Year-Old Beer Re-Created From Chinese Recipe. *National Geographic News*, 2005, Accessed on 27 September 2015, URL: http://news.national-geographic.com/news/2005/07/ 0718_050718_ancientbeer.

35. Robinson, J., (2006). *The Oxford Companion to Wine.* 3rd Ed., Oxford University Press, pp. 230.

36. Rodrigues, F., Caldeira, M., & Câmara, J. S., (2008). Development of a dynamic headspace solid-phase microextraction procedure coupled to GC–qMSD for evaluation the chemical profile in alcoholic beverages. *AnalyticaChimicaActa, 609*(1), 82–104.

37. Sanchez, O. J., & Cardona, C., (2008). A.Trends in biotechnological production of fuel ethanol from different feedstocks. *Bioresource. Technology, 99*, 5270–5295.

38. Shakhashiri, B. Z., (1983). *Chemical Demonstrations: A Handbook for Teachers of Chemistry.* Volume 1, University of Wisconsin Press, pp. 546.

39. Shigechi, H., Fujita, Y., Koh, J., Ueda, M., Fukuda, H., & Kondo, A., (2004). Energy-saving direct ethanol production from low-temperature cooked corn starch using a cell-surface engineered yeast strain codisplayingglucoamylase and a-amylase. *Biochemical Engineering Journal, 18*, 149–153.

40. Singh, B. D., (2009). General and industrial microbiology. In: *Biotechnology Expanding Horizons.* 2ndEdn. Kalyani Publishers, New Delhi, pp. 23–50.

41. Stewart, G. G., & Russel, I., (1987). Control of sugar and carbohydrate metabolism in yeast. In: *Bery, D. R., Russel, I. and Stewart, G. G. Eds.,Yeast Biotechnology.* London: Allen and Unwin, pp. 279–310.

42. Sues, A., Millati, R., Edebo, L., & Taherzadeh, M. J., (2005). Ethanol production from hexoses, pentoses, and dilute-acid hydrolyzate by *Mucorindicus. FEMS Yeast Research, 5*, 669–676.

43. Sukumaran, R. K., Surender, V. J., Sindhu, R., Binod, P., Janu, K. U., & Sajna, K.V.; et al., (2010). Lignocellulosic ethanol in India: prospects, challenges and feedstock availability. *Bioresource Technology, 101*, 4826–4833.

44. Sun, Y., & Cheng, J., (2002). Hydrolysis of lignocellulosic materials for ethanol production: a review. *Bioresource Technology, 83*, 1–11.

45. Taherzadeh, M. J., & Karimi, K., (2008). Enzyme based hydrolysis for ethanol from lignocellulosic materials. *Bioresources, 2*(3), 472–479.

46. Ward, O. P., (1989). *Fermentation Biotechnology principles, processes and products.* New Jersey: Prentice Hall, pp. 321.

47. Windish, W.W., & Mhatre, N.S., (1965). Microbiol amylases. *Advances in Applied Microbiology, 7*,273.

48. Wingren, A., Galbe, M., Zacchi, G., (2003). Techno-Economic evaluation of producing ethanol from softwood: comparison of SSF and SHF and identification of bottlenecks. *Biotechnology Progress, 19*, 1109–1117.

49. Wyman, C. E., (2010). Cellulosic Ethanol: A Unique Sustainable Liquid Transportation Fuel. *Biomass and Biofuels, 33*, 381–382.
50. Wyman, C. E., Dale, B. E., Elander, R. T., Holtzapple, M., Ladisch, M. R., & Lee, Y. Y., (2005). Comparative sugar recovery data from laboratory scale application of leading pretreatment technologies to corn stover. *Bioresource Technology, 96*(18), 2026–2032.

CHAPTER 8

FOOD GRAIN STORAGE STRUCTURES: INTRODUCTION AND APPLICATIONS

VISHAL SINGH, DEEPAK KUMAR VERMA,
and PREM PRAKASH SRIVASTAV

CONTENTS

8.1   INTRODUCTION

Grain is an important consumable commodity that is used for human consumption at large scale. The production of different types of grains increased progressively due to implementation of advance production practices but because of improper storage facilities huge amount of grains has been spoiled. Total 12 to 16 million metric tons of food grains were

In this chapter: One US $ = 60.00 INR.

spoiled due to unavailability of proper post-harvest handling of grains every year which is sufficient to feed the 33% population of India and total yearly grain losses has been estimated near about INR 50,000 crores [46]. Availability of food grain for the consumers can be increased with increasing the advance post-harvest management practices because of world population has been reached nearly 9 billion up to 2025 needed lump some 2.8 billion cereals, 1.6 billion animal products and 5.3 billion of other crops for consumption in the limited area of cultivation which is only possible if all the produced food grain has preserved with minimum losses [23]. In appropriate storage system caused contamination in the food grain [43].

Generally, 10–20% of market price can be enhanced during off-season, which is advantageous for the farmers due to storage of grain with the help of modern storage structure [8]. Losses during storage depends upon several factors like used storage structures, duration and purpose of storage, grain treatments just before storage. Insects, rodents and micro-organisms also hamper the quantity and quality of the stored food grains because the several insect-pests have been affecting the grain during storage and growth of food grain insects-pest also depends upon climatic and place where the grains storage has been placed. According to Sinha and Sinha [48], 10–15% of food grain was lost within single season due to the insects-pest attack during storage and also reduced the food grain quality by spreading the unwanted odors and flavors [48].

Different storage structures generally have been used for maintaining the quality of the stored food grain for long time with minimal microbial and insects-pest losses. Lorini [25] stated that 20% food grain was lost due to improper storage structure and inadequate storage practices and these food grain losses have very high quantitative and monetary value on the world level [39]. Types of storage structures and storage management have very vital role for preserving the agricultural produces but in the some regions of the world still ancient storage structure are used for storing the food grains. The need of food storage arises due to supply the food grains to the consumers in the off-season or time of requirement means food grain storage works as bridge between production and consumption. Different storage structures and methods of storing the food grain also depend upon the socio-economic status of the particular regions [50]. In present scenario, number and capacity of safe and advanced storage structures should be increased with increasing food grain production. Bulk storage of food grain supervised with the help of different government agencies for minimizing the different losses during storage period because stored grains are mostly affected due to insect-pest which reduces the nutritional and other qualities [27].

Sustainable and eco-friendly concern traditional knowledge has been improved by several years of trial and error observation of farmers, technological modification of experts for meeting with the need of a reliable and evergreen farming. Indigenous farming practices which are still in uses widely has been remarked as well as accepted by progressive farmers of those regions or scientifically approved by experts [6, 30]. As we know that a major part of world population is based on agriculture for obtaining their food and the people are involved in farming directly for producing the food known as farm-eINR. In Southern part of Asia like India, Nepal, Bangladesh, Pakistan and Bhutan, etc. have highest number of farmers is falls under marginal and semi marginal category which have less land holdings that is why they need small and cheap agricultural practices to grow and preserved their agricultural produces. So for solving the problems of marginal and semi marginal farmers we have to focus on existing indigenous approaches and its technological improvement [30]. Indigenous technologies have been developed by the local farming communities for harnessing the locally available natural and easily accessible resources to develop the cost effective and eco-friendly structures and practices [6, 13, 37].

Traditional storage structures used in rural community is very useful to keep the stored grain safe by sealing the inlet and outlet after filling the grain in it [21]. Different categories of storage structures like small and large capacity, in-door and outdoor, traditional and modern, temporary and permanent and Government or private were developed in different regions influenced by materials available, grains quality, quantity and types, storage duration, socio-cultural ethics and environmental factors, etc. [14, 17, 20]. In rural areas generally farmers are storing the food grains for their yearly consumption, seed purpose as well as for selling the excess grains to the market [21; 36].

Storage structures are playing key role to prevents the losses and keep the grain safe during storage by minimizing the effect of rodents, micro-organism and environmental factors to feed the growing population of the world [18, 22]. Due to application of several modern technologies several modified and advanced grain storage structures are comes into the picture and widely used by different industries and governments to safely store the grain up to long duration but still advanced storage structures and modern practices has not been available as much it is needed to store the entire surplus food grains of world due to which a huge percentage of total pro-duced food grains (nearly 10–20%) of world were going to waste every year [22, 47].

Production of food grains and its marketing of surplus quantity after consumption is a one of the main source of income to the farm-ing communities indicates that the high grain production directly influ-enced the socio-economic values of the rural areas. Farmers are going to store generally 59–70% of overall production for their consumption and surplus amount has been stored into small storage structures made with readily available materials in local because they have to store small amount of grain only due to their small land holdings as well as cannot afford the price and maintenance of advance and big storage structures [42].

This chapter discusses applications and potential of food grain storage structures.

8.2 ADEQUATE ASPECTS OF STORAGE STRUCTURES

Storage of different food grains required scientific and modern approach because all the grains are content biological activities and respires that is why rapidly affected due to several insect-pest. Intensive studies of destruc-tive insect-pest (insects, micro-organism, rodents, birds and squirrel, etc.) and factors (moisture, relative humidity (RH), temperature and location of storage, etc.) which provides the favorable conditions to the growth of these destructive agents are necessary. Effect of micro-organism and insects-pest can be controlled up to certain extent with development of adequate condition within the storage structures and most suitable temperature was considered 21°C and moisture level nearly 9% within the bin [27].

8.2.1 DRY AND COOL STORAGE

Pallets made by either wood laths or poles used for storing the food grain bags in form of stacks. Proper arrangement of stacks is important for adequate air circulation throughout the stacks for maintaining the moisture content and temperature of the food grains and for preventing the stored food grains from water, pallets has been covered with plastic sheets [12].

8.2.2 AIR TIGHT STORAGE

Air tight storage systems used for storing the food grains in absence of oxygen because of insects-pests cannot survive in the storage structure without oxygen. Containers with tight fitting lid (like basket made by bamboo and layered with clay, plastic, metals, etc.) or underground storage structures have been used for storing the grains because un availability of air retard the micro-organism, insect and pest growth completely and enhanced the storage life of the food grains [12].

8.2.3 HERMETIC STORAGE TECHNOLOGY

Hermetic storage protected the stored food grains from rodents, insect-pest, and moisture migration by creating suitable storing atmosphere within the storage structure with the help of modified atmosphere (less oxygen and high carbon-dioxide availability). Research has been revealed that less expensive conventional storage structures are suitable for less duration of storage (nearly 3 months) further this structures are not adequate for long storage due to microbial and insect-pest attack. Hermetic storage has been efficient to store the food grain without hampering the quality aspects up to long duration [40].

8.3 STORAGE STRUCTURES

Throughout Asia, farmers are going to store nearly tree forth of the total grain production for food, seed and cattle feed as well as nearly one-fourth part is selling into market. Food grains have been stored in different traditional and modern storage structure for avoiding the quality degradation of food grains. Generally grain storage structures are classified into following types according to available materials, economical aspects and storage capacity, etc.

- Indoor and outdoor storage structures
- Above-ground and under-ground
- Rooms/bins/pots constructed with mud
- Wood or Bamboo storage structures
- Metallic drums, bins or containers
- Structures made with straw of paddy and wheat, etc.

8.3.1. DOMESTIC STORAGE STRUCTURES

Domestic structures are generally used for small storage capacity (like guards, small baskets, tin containers, baskets, earthen pots and jute bags) manufactured with the help of wood, metals, bamboo, mud and plastic (Figure 8.1). Storage structures should be properly covered and should prevent the air and moisture migration to avoid oxygen availability which slows down the microbial activities during storage [28]. Indigenous storage structures are basically used for strong the food grain in rural areas on small-scale. In India, 59–70% of food grains is stored in the traditional storage systems, which are constructed with the help of paddy or wheat straw, bamboo, wood, bricks, mud, cow-dung, etc. storage of grains can takes place inside, outside the houses or underground storage structures [5]. Indoor storage structures like Kanaja, Kothi, Sanduka and earthen pots of different sizes have been used for small quantity of grain storage. Some other storage structures were constructed underground as well as above the ground like "Khatties" also called "banda" considered as underground storage structure provides better control of moisture and temperature and facilitate airtight condition to the stored food grain. Different storage

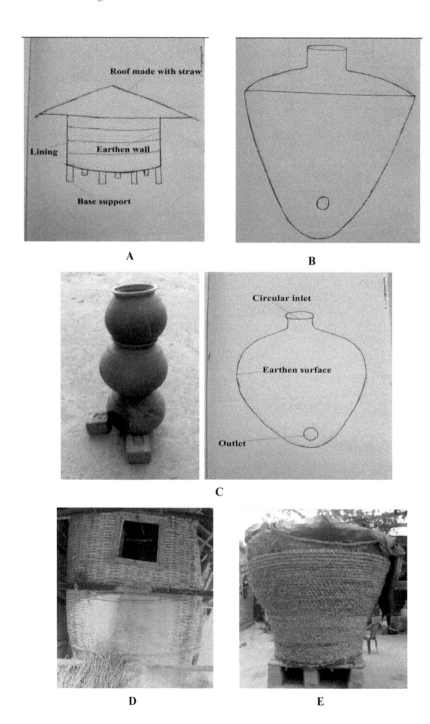

A

B

C

D   E

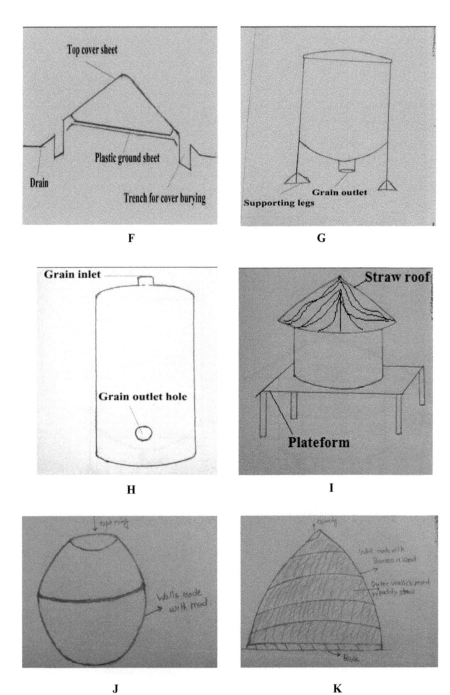

F

G

H

I

J

K

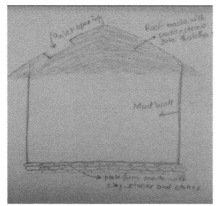

L

M

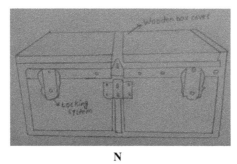

N

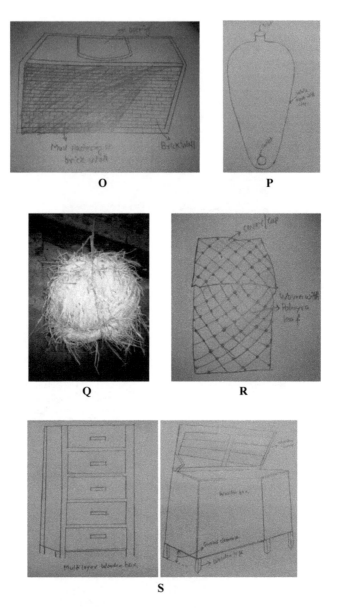

FIGURE 8.1   Different domestic food grain storage structures:
(A) Kanaja, (B) Kothi, (C) Earthen pot, (D) Gummi, (E) Hagevu, (F) Pit, (G) Bukhari,
(H) Mud bin, (I) Traditional storage structures, (J) Bharola, (K) Kupp, (L) Kodambae,
(M) Bamboo basket (Urai/Peru), (N) Wooden box (Sanduk/Arisi petti), (O) Mud house
(Mankattai), (P) Earthen bin (Kulukkai), (Q) Straw bin (Thallpai), (R) Palmyra leaf bin
(Vattappetti), and (S) Wooden bin (Mara thombai/Pathayam).

structures (like Kothi, Kuthar, Kuthla, Kanaja, Thekka, Puri and Morai, etc.) are constructed above ground with the use of mud, bamboo, bricks, straw, etc. Storage structure above the ground facilitates easy inspection of the food grain, turning for avoiding the hot spot [27]. Traditional storage structures are economical; and less construction and maintenance skills are required; and these have small capacity and used for holding the grain for less duration (like traditional crib, underground storage and barn, etc.) [28]. Different traditional storage structure developed by farming communities of different regions are based on their indigenous skills, ethics and available materials [49].

8.3.1.1   Kanaja

Kanaja is a traditional storage structures basically made with bamboo and plastered with either mixture of cow-dung and soil and covered by paddy straw (Figure 1A). Generally it has round base and large inlet for grain filling at the top and after filled up the grain, inlet also sealed with mixture of cow-dung and soil. Small opening has been provided in the lid for taking out the grain without opening the sealed lid. Height and capacity of the storage structure varies and 3–12 quintals of grain can be stored. More than one types of grain also can be stored in Kanaja, which has two or more partitions. The storage system balanced on several stands or on foundation made with wooden plank, stones and bricks, etc., which height kept nearly 12 inches above the ground level for preventing the migration of soil moisture into storage grains [31]. It is also called "Thombai" in Tamilnadu, where its outer and inner side were plastered with help of mixture of cow dung and clay and let it leave for drying in sun. After drying, top portion of structure has been covered with crop residues, long grasses, straw and leafs of locally available trees, etc. in conical shape for avoiding the enter of rain water into the structure. Generally ginger grasses are used to prevent it from rain [24].

8.3.1.2   Kothi

Kothi is a traditional, small capacity and widely used structure different rural community of India, consist wide door for pouring the food grain inside the structure and small outlet is provided for taking out the grains (Figure 1B). Generally, it is used for storing the jowar and paddy [31]. It is constructed in cylindrical shape in different dimensions for storing the grains in the range of 1 to 50 tons and commonly used materials are un-burnt clay, straw and cow dung [6].

8.3.1.3   Earthen Pots

It is used in the rural areas of South Asian countries like India, Bangla-desh, Pakistan and Nepal on wide scale for storing the small amount of food grain for short duration of storage period. Size and numbers of earthen pots can be increased or decreased according to amount of grain have to store in one pile. Mouth of all the pots has been sealed with help of mud, cow dung or its mixture for avoiding the moisture or air migra-tion. Earthen pots made with burnt clay in several shape and sizes (Figure 8.1C), according to requirement were generally used for keeping small amount of grains inside the houses.

Earthen pots have been kept one above the other vertically started from ground level, this arrangement of this storage structure also called as dokal [5]. It is also called "Mataka" in Uttar Pradesh, Bihar, Uttarakhand, Jharkhand and Madhya Pradesh. In southern part of India also it is known as "Paanai" or "Addukkupaanai" used as single pot for storing the small amount and piles of earthen pots arranged vertically one above another (largest size earthen pot should be on bottom) for more amounts of grains [21, 24].

8.3.1.4   Gummi/Kuthar

Gummi is circular or hexagonal in shape consider as outdoor type stor-age structure (Figure 1D). Generally, constructed with the help of bam-boo strips or reeds and plastered with the mud after final construction for preventing the linkages. Entire structure was placed approximately one meter height from the ground for preventing the moisture movement from the ground towards the stored grains and base of the structure is made by using reeds or stone slabs [5]

8.3.1.5 Hagevu

Hagevu is generally considered as outside storage structure made with the paddy straw rope by arranging the straw ropes one above the other in rounded shape and sometimes also constructed as indoor structure for storing less quantity of food grain (Figure 8.1E). After completely filled up the storage structure covered with thick layer of straw and plastering with using mud for proper air sealing and inlet opening has provided at some particular height from the bottom. Hagevu is treated as short term food grain storage structure but not suitable for seed storage and suitable for dry regions [5]. For insuring the safe storage of food grains inside regular mud plastering for proper sealing has been required [31].

8.3.1.6 Barn, Shelf and Pit

Barn, shelf and pit are the storage structures (Figure 1F), frequently used for protecting the food grains from the un-favorable environmental conditions. Underground structures like pits are properly sealed with lining of nylon or straw and able to store the grains up to long duration without any major qualitative deterioration [28].

8.3.1.7 Bukhari

Generally mud, bricks and cement was used for constructing the bukhari in which space is provided in the bottom portion for easy taken out the food grains (Figure 8.1G). Generally it is constructed in square shape [21]. Polythene sheets provides moisture protection and mud has been used for proper sealing. Some intensive studies reveal that farmers are providing a layer of sand before storing the wheat grains which act as primary barrier to protect the food grains from insects-pest up to some extent [6]. Timber or bamboo strips were used for constructing of walls and floor and generally available in cylindrical shape with conical roof which is made of straw or bamboo. Capacity of food grain storage varies within the range of 3.5 to 18 tones. Now days improved bukhari structure also used frequently for providing better protection to the stored food grains. Improved structures has been prepared with double layer of bamboo strips and gap between two bamboo strip layers was filled up with mud for strengthen the structure and the conical shape roof plastered with 4–5 cm thick mud layer due to protecting the leakage during rainy season [41].

8.3.1.8   Bharola

Bharola is a traditional storage structure easily transportable due to light in weight and small capacity (40–80 kg) made with mud [6]. It is widely used in Punjab, Haryana and some other northern part of India and Sindh and Punjab region of Pakistan also for storing the food grains for daily ways consumption, small quantity for sailing as well as seed also can be store in it. Commonly constructed in round shape with one round opening at top of structure which can be used for both pouring and removing the stored grain (Figure 8.1J).

8.3.1.9   Kupp

Kupp is an indigenous storage structure used for storing the food grains in agricultural farms itself. It is designed in conical shape with wide circular base for better stability. Bamboo splits or wooden thin sticks were attached with the help of rope made with jute or paddy straw to form a round base and conical top (Figure 8.1K). It is temporary storage for short period. Kupp having only one opening at top which should be covered with thick layer of paddy straw, crop residues, bamboo splits and cotton sticks for avoiding the entry of rain, direct sun light and air [6].

8.3.1.9.1   Kodambae

It is a traditional storage structure frequently used in South Asian regions, mainly in India (Tamilnadu, southern part of Andhra Pradesh and Telangana), Bangladesh and Pakistan for storage of food grains, seed and other fodder crops. Locally available materials like soil, clay and mud by mixing cow dung with soil were used for constructing a round wall up to 3 feet height on nearly 0.5–0.7 feet high platform from ground (made with mixture of soil and clay, stones, bricks, etc.) and after that its covered in conical roof made with natural materials like paddy straw, palm leaves and wooden or bamboo sticks stitches with jute or paddy straw rope, etc. Sometime bricks, cement, stones and sand also used to construct a wall and base when for strengthening the storage structure [21]. Generally grain holding capacity is noticed nearly 1 tons but it may vary as per requirement. For loading and unloading the grains in this structure one small door has been provided through which easily one person can go inside with help of ladder (Figure 8.1L).

8.3.1.9.2 Bamboo Basket

Bamboo baskets are used for grain storage from long time back in rural areas. It is used throughout Southern part of Asia and its size varies according to quantity of grain have to store but generally classified under inside, temporary (can be used up to 4–5 years), small capacity and short duration storage (can store the grain up to one year without degradation in quality). Bamboo splits has been stitches in the shape of basket. One circular opening is provided at top for storing and taking out the grain and during storage it's covered with flat cover made with bamboo or wooden sticks (Figure 8.1M). After filling the grain, this structure were completely layered with mixture of cow-dung and soil for sealing the small wholes which make it hermetic and protects from rodents, insect and pest attack. It is known as "Urai" in South India [21]. Similar type of structures was also used in Himachal Pradesh region with the name of "Peru" constructed by stitching the bamboo strips (width – 2 cm). In Himachal Pradesh, it is kept on a platform made with locally available wooden plank called as "Tarein" to avoid the micro-organism and insects attack [19, 20].

8.3.1.9.3 Wooden Box (Sanduk/Arisi Petti)

Wooden box is very popular traditional, small capacity, inside storage structure used in different parts of Southern Asia regions like India (Uttar Pradesh, Uttarakhand, Bihar, Rajasthan, Jharkhand, Madhya Pradesh,

Chhattisgarh, Andhra Pradesh, Karnataka and Tamilnadu, etc.). It is generally made with wood of strong and locally available trees in rectangular or square in shape. Its dimension depends on the quantity of grain have to store (Figure 8.1N). It is also called by different names in different regions like Sanduk or Peti in North India region, Pakistan, Bangladesh and Arisi petti in Southern region of India. Its have lock system also to protect the grains from thieves. This type of storage structure was very suitable to avoid the attack of weevils, insects, rodents and moisture migration [21]. It is called as "Peti", particularly in Himachal Pradesh region, where it is made with preferable particular wood called "Tuni" (*Cedrela toona Roxb.*) and Akhrot (*Juglans regia Linn.*) which have less effect of termite and other insects [19].

8.3.1.9.4   Mud House

It is a traditional storage structure practiced from long time in rural areas of India and some other parts of Southern Asia region for grain storage. It is cost effective categorized under inside, temporary and small capacity of storage structure (Figure 8.1-O). In this structure, the walls constructed with earthen bricks and plastered with cow dung and after that it's covered with board prepared with wooden or bamboo and also called as "Mankattai" in Tamilnadu [24].

8.3.1.9.5   Earthen Bin

The traditional storage structure "Earthen bin" is used for grain storage up to 2 quintal and kept inside the house. It is cylindrical in shape with narrow bottom and wider upper portion with a narrow neck like opening for pouring the grain and small opening at bottom for grain removing. After pouring grain, top and bottom opening should be closed with help of mud, clothes and straw, etc. It is successfully used for storing the paddy, black gram and millets and also known as "Kulukkai" (height and diameter is 2 and 0.5 meter, respectively) in Tamil Nadu [24]. Size of this structure varies according to requirement (Figure 8.1P).

8.3.1.9.6 Straw Bin

Straw bin is a small capacity indigenous storage structure made with straws (generally with paddy straw) for storing the grain or seed. Long twisted rope made with paddy straw was arranged in circular shape on ground (long rope arranged in same structure like one ring kept on other ring up to a certain height). A thick base of structure is made with locally available crop residues like leaves and hatches of palm, jute and coconut trees, long grasses and paddy straw, etc. This is categorized as temporary and low cost storage structure popular in the rural areas of different Asian countries like India, Pakistan, Bangladesh and Nepal (Figure 8.1Q). It is also called as "Thallpai" in Tamil Nadu region, where generally it is constructed in small size for storing the seed and kept by suspending with the roof [24].

8.3.1.9.7 Palmyra Leaf Bin (Vattappetti)

The traditional storage structure made with Palmyra leaf is widely used in Southern regions of India for short term grain storage. It is generally made by stitching the Palmyra leaves and appears in cylindrical shape contains storage portion and its cover (Figure 1R). Its dimensions can be vary and mainly depends on the requirement of farmers but common dimensions were observed as height, width and length are 2.5–3, 1 and 2 meter, respectively which can store grain up to five tons [24].

8.3.1.9.8 Wooden Bin (Mara thombai/ Pathayam)

Wooden bin or boxes were rectangular in shape and commonly used in different rural regions of India, Pakistan and Bangladesh as inside storage. Small and large capacity rectangular wooden boxes were known as wooden boxes and wooden bin, respectively. This type of storage has been constructed in different dimension according to recommended storage capacity. It contains single box or different sections (one rectangular box fitted above another like a drawer), which can be open and closed as drawer. It should be painted with good quality paint or primer for increasing the longevity of storage structure and avoiding the attack of termite, insects and pest (Figure 1-S).

In Himachal Pradesh and other northern region of India, it is known as "Kuthar", generally made with Deodar wood and can be used up to 70–80 years for grain storage [19]. Wooden box or bin also known as "Mara Thombai" in Tamilnadu, where it can store nearly 1 tons of grain in the structure which have 2.13 and 1.21 meter height and width, respectively [21]. In southern most regions of India like Kerala, southern part of Karnataka and Tamilnadu similar wooden box or bin known as "Pathayam" were prepared with help of light wooden planks but it contains only one opening at the top which is covered temporarily with thatches or leaves of different trees [24].

8.3.2 MODERN STORAGE STRUCTURES

Due to expansion of agricultural land and used of several technologies and modern practices, the food grain production is increasing gradually. So in modern age for handling and preserving the huge amount of grain up to long duration many modern grain storage structure has been introduced which is generally used by different grain processing industries, food corporation of India, Grain mills and whole seller. Prevention of significant qualitative and quantitative losses the traditional storage structures were improved technically and appear more efficient for storing the food grains. Different advanced storage structures like ware house, silo, bin, cold storage, hermetic storage system and controlled atmospheric storage system (Figure 8.2) were used for storing the food grain according to availability, suitability and acceptability for particular types of food grains, durability of storage and place of storage. Traditional storage structures has been progressively modified and improved for better storage of grain for long time without any spoilage known as modern storage structures. The storage structures like Pusa bin, PAU bin and Hapur tekka has been used for small quantity storage. Bins are used for food grains storage generally placed inside the houses [3].

8.3.2.1 PAU Bin

Punjab agricultural university has been designed a storage structure called as PAU bin, constructed with galvanized iron sheet. Capacity varies between 1.5 to 15 quintals depends upon the size of designed PAU bin [1].

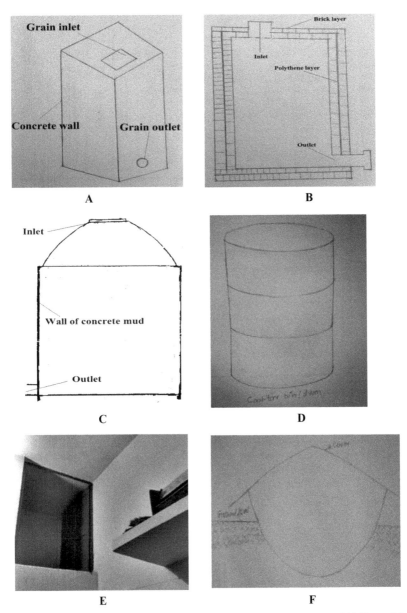

FIGURE 8.2   Different modern food grain storage structures: (A) Pusa bin (outer structure), (B) Pusa bin (rectangular cross section), (C) Modified Domestic bricks or cement bins, (D) Coal-tar drum or bin, (E) Granary room , (F) Semi Underground Cement Bin, and (G) Cover and Plinth Storage (CAPS) Structure, (H) Silo, (I) Warehouse, and (J) Hermetic Storage.

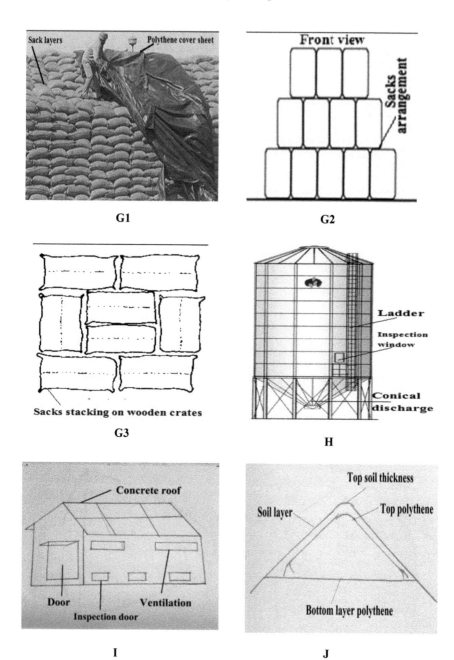

FIGURE 8.2   *(Continued)*

8.3.2.2   Pusa Bin

Pusa bin is considered as modern small-scale storage structure designed by constructed with the help of mud or bricks and polythene sheets used for providing proper sealing within the wall of the structure (depicted in Figures 8.2A and 8.2B). Concept and design of Pusa bin which constructed with the help of mud or bricks has been proposed from Indian agricultural research institute. Polythene sheets was placed within the bricks or mud layers for providing proper sealing which act as moisture barrier [1]. Pusa bin was constructed on the mud bricks platform which covered with plastic sheet of 700 gauges for protecting the stored grain properly that is why it is also known as "low density polyethylene" storage structure [41, 42].

8.3.2.3   Hapur Tekka

Hapur tekka is cylindrical in shape, locally constructed on base of metal tube with the help of bamboo and expandable clothes and another hand in improved structure, bamboos has been replaced with galvanized iron or aluminum sheets for getting more durable structure. Grain has been taken out with the small circular or rectangular outlet provided in the bottom of the structure. Indian grain storage institute existed in Hapur have designed Hapur tekka for storing the food grain safely [1]. Generally Hapur tekka can contain grains of 2 to 10 tons [42].

8.3.2.4   Modified Domestic Bricks or Cement Bins

Modern Clay bins have cylindrical shape (Figure 8.2C), frequently used for storing the food grains in domestic purpose and for manufacturing with materials like burnt bricks, concrete and alkathene sheet, etc. has been used for making a strengthen, moisture and air proof structure due to which food grain can be stored for certain period of time without any qualitative deterioration [27].

8.3.2.5   Coal-Tar Drum or Bin

Central Institute of Agricultural Engineering, India has developed a grain storage structure made with coal-tar (Figure 8.2D). It is categorized under low cost, small storage capacity and durable storage structures [42].

8.3.2.6   Cement Wall Bin

A rectangular shape storage structures in which cement, bricks, sand and stone has been used for constructing the side walls as well as base. After constructing the structure, inside and outside walls are plastered properly to obtained smooth wall surface which avoid the movement of rodents and termite action. Most adopted dimension of this structure is 2.43*1.82*2 m (length, breadth and height, respectively) but it varies, if higher amount of grain have to store [21]. It is a permanent type of storage structure used for long term storage of food grain. This type of structures widely used in Southern part of India and also called as Kalangiyam in Tamilnadu.

8.3.2.7   Granary Room

Roof almirah is considered as inside, permanent and modern storage structure constructed with the help of concrete, bricks, sand and cement inside the houses (in rooms, kitchen and store rooms, etc.). Roof almirah is designed by providing a concrete or bricks slab (RBC or RCC slab) just above the doors of rooms or in corner of wall bellow the roof of house with one mini door for easy in grain handling [21]. Minimum effect of rodents was observed in this type of structure because grains have been stored at height (nearly 6.5 to 7.5 feet height from house floor). Grain stored by either spreading it on floor of storage or kept by filling into the bags and ladder is used to poured or collect the grains. It is constructed in all over India and called as "Du-chaatti or Kotha" in

Uttarpradesh, Bihar and Uttarakhand as well as Macchu in Tamilnadu (Figure 8.2E).

8.3.2.8   Semi Underground Cement Bin

Semi underground cement bin is semi-spherical in shape and nearly half portion existed bellow the ground level and half above the ground used for storing the food grains (Figure 8.2F). Semi-spherical bottom shape of structure placed bellow the ground provide strong support to the bin and main advantage that bin is considered as cast-in-situ does not required any extra floor space in room [27].

8.3.2.9   Cover and Plinth storage (CAPS) Structure

Large quantity of food grains was stored with the use of large scale storage structure. Cover and plinth storage (CAPS) structure is constructed with the help of wooden crates which fixed into the grooves of brick pillar and generally first stack has been taken place nearly 1.17 feet height above the ground (Figure 8.2G-1). The stack arranged one above another on the wooden crates in vertical manner and finally covered with 250 micron low-density polyethylene (LDPE) sheets from the all sides (Figure 8.2G-2, G-3). CAPS structure is very useful for storing the different types of food grains like cereals, pulses and millets up to one year [21]. Food Corporation of India (FCI) is frequently used this storage structure for storing the bagged grain because of it is most economical and can be constructed within 3 weeks. Large amount of grain can be stored in this structure economically [31]. Cover of this storage structure is rectangular in shape and made of polythene sheet which has been open from lower side used for covering the arranged sacks [41].

8.3.3   OTHER STRUCTURES

8.3.3.1   Silo

Silos are constructed either with metal or concrete but according to the economical point of view metal silos are better than the concrete silos

(Figure 8.2H). Capacity of the silos depends upon the diameter and height of the silos. Approximately 25 thousand tons of food grain can be stored in this storage structure and loading and unloading in the case of huge silos performed with the help of elevators and conveyor Silos are generally constructed with metals, aluminum, concrete and rubbers used for long term storage of bulk quantity of food grains. Silos are considered as advanced structure which required modern techniques and accessories for maintaining the food grain qualities during storage. Generally Silo or bins are cylindrical in middle and top portion is conical in which augur rotates for aerate the grains and prevents the chance of hot spot creation [28]. Different insects which responsible for spoilage of food grains cannot survive below certain temperature (16°C) and temperature within the range of 16–21°C is considered as suitable range which retards the activities and multiplication of most of all insects and stored grain spoilage is non-significant at lower temperature [10]. Moisture and RH is two important factors need to be controlled for long storage. Navarro and Noyes [32] stated that Moisture content and temperature of storage is 12.5% and 25°C, respectively provides most favorable condition for long term food grain storage without any major losses.

Generally first time construction of metal and concrete silos started in Europe and United States because these types of storage structures were suited the climatic condition of temperate regions because of proper ventilation arrangements is provided for cooling the grains during low temperature season but these type of structure was not advised in warm climatic regions for avoiding the condensation and humidity effect which caused the destruction of stored grains due to micro-organism attack [7]. Traditional storage structures most commonly used in Sub-Saharan region for storage the food grains, not able to preserve the stored food grains efficiently via complete destruction of the insects-pests. Kimatu et al. [23] found that sometimes weevils caused more than 30% of total post-harvest losses in the case of traditional storage structure was used that is why the Food and Agriculture Organization (FAO) have initiate to construct the metal silos which can protect the stored food grain from insects-pests and storage time also can be enhanced. Metal silos are efficient to keep the food grain up to long duration with controlled respiration, moisture content, etc. and facilitate the unfavorable conditions for several insects-pests growth [16].

8.3.3.2 Warehouse

Warehouses are used for storing the bulk amount of food grain for long duration in scientific way. Food grain can be properly stored in the warehouses without any significant qualitative and quantitative losses. Different government organizations like Central warehouse corporation, State warehousing corporation Food corporation of India, and industries has constructed own warehouses for storing the heavy quantities of food grain for long period of time. Warehouses can store the food grains either in bags or in bulk. It can store heavy amount of grain under one roof which provides the facility to transport or export collective quantity of food grain when required [31]. Warehouse constructed with bricks, concrete-cement and metals and height of foundation floor always kept nearly 0.45 to 1 meter from ground level which prevents the frequent entry of rodents. Within the structure, wooden pallets are used for stalking the sacks of grains which prevent the moisture migration from floor to food grain. Adequate facilities of protection like proper sealing of the doors, maintenance of inspection doors, and ventilation controlled, etc. has been very necessary. Proper stack formation with maintaining the standard distance between two consecutive stacks (nearly 1.7 meter^3 for 1 ton of food grains) is important for easy handling operations (Figure 8.2I).

8.3.3.2.1 Central Warehousing Corporation

For storing the bulk amount of harvested food grain, the government of India have founded Central warehousing corporation in 1956 which provides the storage facilities to 120 agricultural produces with adequate quality and hygienic parameters [31]. According to the Indian government, Central warehousing corporation is the largest public controlled food grain storage system throughout the country and also provide facilities like cleaning of warehouse area, handling, transportation, procurement, distribution, insect-pest disinfection and other quality control aspects.

8.3.3.2.2   State Warehousing Corporations

State warehousing corporation has been constructed state wise for storing the food grains of particular states. Different district of Indian state are facilitated with state ware housing for bulk storage of food grain of particular area and stored amount of grains of any particular states warehouse accounted as share of that state at India level.

8.3.3.3   Food Corporation of India (FCI)

Food Corporation of India (FCI) is constructing a large capacity storage structures also called as godowns, has a largest food grain storage capacity near about 26.61 million tons in India. For adequate storage of grains, the site selection of storage, cleaning and fumigation facilities and inspection of grain quality after certain intervals have very important role. Most common pest responsible for damaging the stored food grain is *Callosobrunchus* spp, *Trogoderma granarium*, *Tribolium confusum* like beetles, *Acanthoscel idesobtectus* (weevils), *Corcyra cephalonica* (moth) and rodents. Several factors like temperature, moisture content of the stored grain, adequate fumigation, duration of storage, etc. are responsible for growth of different pest. Several treatments like prophylactic Malathion (emulsifiable concentration is 50%), Dimethyl 2, 2-dichlorovinyl phosphate (emulsifiable concentration is 76%) and Deltamethrin (2.5% WP) has been implemented for controlling the attack of pest and storing the food grain for long duration as well as fumigant like aluminum phosphide is used in air tight storage structure for controlling the grain infestation [21, 31].

8.3.3.4   Hermetic Storage

Oxygen level was lowered down up to 1–2% due increasing the carbon dioxide level or creating vacuum [9, 56]. Hermetic storage becomes very popular in several countries like Ghana, Philippines, Rawanda, Srilanka and Sudan. Large hermetic storage called as Bunkers commonly used from last decades of 20th century and constructed with different capacity range from 10 -30 thousand tons (Figure 8.2J). Donahaye et al. [7] stated that hermetic bunkers have been suitable for storing the wheat of 12.5% moisture content for four years without any qualitative degradation [7]. Hermetic storage structures are frequently used for storing the food grains and provide easy portability, avoid the insect-pest infestation and retain the qualitative characteristics during storage period. Hermetic storage containers are very efficient to protect the food grain and reduces the losses up to 0.15% as well as increased the shelf life with the destruction of mould and other micro-organism (up to 99.9%) because of this storage technology has been create air tight and sealed as well as creates carbon-die-oxide enrich and oxygen reduced environment [33, 51, 55]. In this technology, stored grains have no need any chemical treatments and fumigation because of healthy storing condition has been developed with the help of modified atmosphere method [52]. Construction of hermetic storage structures depends upon the amount of grain which have to stored that is why the capacity range of the structure is varies between 0.59–1 tons known as Super grain bags, 5–30 thousand tons grains can be stored in the hermetic storage structure called bunkers and Cocoons [53].

8.4   LOSSES DURING STORAGE

Stored grains in storage structures were spoiled due to attack of different insects-pests, fungi, rats, mice, squirrel, birds, etc. Improved and advanced storage structures efficiently minimize the food grain losses and enlarge the shelf life of food grain without the quality deterioration [12]. Losses of the grains have been taken place due several reasons which interferes and degraded the food grain qualities in the absence of advance storage structures. Storage structures should incorporated the controlled, safe and hygienic storage conditions for maintaining the high quality of the grains which is very necessary to fulfillment of market demand of healthy food grain supply [39]. Adequate atmosphere should be necessary within the storage structures preventing the stored food grain degradation because only insects damaged the 5–10% of stored grains world-wide and these losses extended up to 30% in the tropics area of United States for which annual monetary value is estimated nearly at 200 million dollars. Different insects and spoiled grain provide suitable conditions for growth of fungi and further spoiled the grain [4]. Insects are generate the odor smell of stored grain due to creation of unfavorable condition within the storage structures like development of hot spot, which suitable for fungus development and finally deteriorate the food grains. For supplying the standard food grains in Indian and foreign markets, qualitative parameters should be followed [54]. Major part of post-harvest food grain losses occurred due to weevils attack during storage of grain in an adequate traditional storage structure which is not efficient to prevent the insect-pest affect completely [15]. Moisture migration has been taken place due to the existence of temperature gradients because of upper layer of stored grains become cooler than internal layers of grains in cold season (atmospheric temperature is lower). Within the storage structure, the air with low temperature has higher density that is why moved along side walls but hot moistened air transferred up to upper layers of cooler grain because of low density which condensed after striking with grains of low temperature. Condensed moisture generate wet-spot is favors the unwanted insect-pest growth. Proper cleaning around the storage structures has been necessary for protecting the grain from several insect-pests because scattered grains facilitate the growth of insects and pest [38]. For reduction of destructive agents during storage of grains has been required application of fumigants, different chemicals and adequate aeration which limitation depends upon quantity and types of the food grains [11]. Commonly food grains are stored after harvesting, cleaning and drying in the different storage structures for certain period of time (long or short duration) within the storage duration different factors like insect-pest, rodents, location of storage, temperature of stored grain and atmosphere, moisture content, water activity and respiration rate has to be controlled with the implementation of suitable management practices under which proper inspection of several critical factors like micro-organism (bacteria, mould and yeast), gas composition within the structure, mycotoxins development, etc. has been essential because several critical micro-organism are active and respiring with slow rate in dry and safe storage atmosphere [26]. Aldred and Magan [2] have observed the adequate and effective management required to develop the strategies to control the food grain losses due to the mycotoxins affect. Infestation of food grain viva insect-pest, rodents and micro-organism enhanced the respiration rate which generates excess temperature responsible for appearance of hot spot during storage, increment in moisture content and carbon dioxide level. Temperature and moisture content rising due to respiration also hampered the quality of food grains above the general recommended temperature and moisture content of storage 15°C and 13–14%, respectively. Quality control and hygiene parameters consideration has been regulated in the food industries for producing safe and nutritious food product without any significant losses with the help of hazard analysis critical control point (HACCP) [29].

8.5 FACTORS AFFECTING STORAGE OF FOOD GRAINS

Several factors are responsible for quantitative and qualitative degradation of the food grains during storage like insect-pest, micro-organism, rodents and birds which reduced the food grain shelf life in same time we have to keep attention on different factors like nature of storage structure, type of commodity stored, duration of storage and quantity of stored food grain as well as climatic factors for keeping safe the food grain during storage [42]. Different chemical pesticide and insecticide used during grain storage also has been affect the quality of food grains that is why now a day's conventional, organic and indigenous pesticide, insecticide were preferred to protect the grain and make it safe for consumption [35]. For maintaining the quality and hygiene during storage, routine inspection, fumigation and plant based natural additives were used [24, 34, 44]. Several new technologies and instruments like "Probe trap" (it is attached with electronic device and computer which counts the number of insects), speed box (containing heater and ventilator used for regulating the fumigation and study the condition of insects) has been developed to study and control of insect's growth [18].

8.5.1 MICRO-ORGANISMS

Micro-organism (bacteria, yeast and fungi) have significant role in damage of stored food grains because growth of several micro-organism caused odor flavor, hot-spot, grain clogging, moisture enhancement due to high respiration, etc. Fungi behaves like parasite within the stored grain caused the severe disease to the consumers and due to growth of fungi the respiration rate of food grain within storage increased gradually and create hot-spot which hampered the good milling characteristics of stored grains. Different mycotoxin also has been produced due to multiplication of fungi which treated as highly toxic for consumers [28]. Spores of fungus are very light and small can be spread one to another places with the help of wind and insects very difficult to removed completely from food grain storage. Fungus spores are able to reproduce themselves and caused the spoilage of stored food grains and deteriorated the quality also. Humid and hot atmospheric condition is favorable for fungus growth sometimes mould growth also takes place at lower temperature and high air RH. Dry weather retards the growth of fungi but did not able to destroy the spores due to high resisting capacity in dry conditions [12].

8.5.2 INSECTS

Growth of insects in the different storage structures depends upon the favorable conditions like moisture, temperature, air, etc. That is why for minimizing the multiplication of insects the storage structure should be properly regulate the moisture, air and temperature within the storage and maintained the adequate condition for preserving the food grain for long time without affecting the shelf life of food grain. Beetles and moths are two major insects affect the different stored food grain and pulses [12]. Insect-pests grow very rapidly if storage structures in which food grain are stored unable destruct the insect-pest effect. During storage period, insect infestation occurred because of un-adequate storage facilities. Insect multiplication takes place because of insect damaged the stored grains and laying the eggs inside which spread out the odor smell, reduces the weight, quality aspect like nutritional value and viability and raised the temperature. Weevils has been consider as most destructive grain insect which spoiled the food grains due to damage the upper layer and hamper the internal edible parts of the food grain.

8.5.3 BIRDS, PEST AND RODENTS

Birds are mainly carrying the food grains for consumption during which minor part of un-covered grains has been infected and facilitate the micro-organism growth but quantity losses has been significant. Rodents are extremely disastrous for storage structures because it makes holes in structures which caused facilitate the easy stored grain infestation and rodents also transported several diseases by means of decay of died one and excreta that is why during construction of storage structure some factors should be considered like hygienic surrounding, proper height of storage structures base and adequate fumigation [28]. Rodents are responsible for damaging the significant quantity of stored food grains by consuming the food grains, damage the grains with their droppings and exposed the food grain because rodents makes several holes in sacks and spread on the floor. It's also spread hazardous disease due to consumption of rodents contaminated food grains [12]. Generally infestation of Insect, birds, rodents affected the quantity losses due to changes in temperature and moisture content as well as qualitative losses affected with reduction in nutritional value, sensory characteristics during storage. Local, cheap and easy available materials used for constructing the grain storage structure in Indian scenario but advance modern storage structure required for long storage without degradation in food grain quality. In rural areas, the small-scale storage structure (like coal tar drum bin, Hapur bin, Pusa bin and chittore stone bin) made mostly with bamboo, mud, stones and wood however stored food grain losses observed around 6% due to un-adequate storage structure. Before storage of the grain, moisture content should be maintained. Large amount of food grain has stored for trading, public distribution, etc. in FCI, State and central warehousing corporation [45].

8.6 SUMMARY

Adequate and efficient storage structures are required to minimize the losses during grain storage because consumption capacity has been increasing with rapid growth of world population. Mechanized high yielding practices as well reducing the food grain losses up to minimum extent has been essential for accomplishment of supplying the food grain as per capita of food grains amount required. Among the total post-harvest losses of food grains, 10–15% losses are due to unavailability of advance and appropriate storage structures, which contribute huge grain deficit that is why for increasing the per capita food grain availability and better market price a hygienic, economical and scientific storage structure has been needed. Advance technologies and intensive studies has a eminent role in development of efficient storage structures which can retard the microbial and insect-pest growth and maintained the quality of stored food grain with enlarged shelf life. An improved and sustainable storage structure is required to control the grain infestation and safe storage of food grain for long duration.

In a given period of time, nearly 59–70% of the total grains were stored in India in traditional storage structures like Earthen pots, Gummi, Kacheri, Kanaja, Kothi and Sanduka for which storage capacity as well as durability of storage is less than the modern storage structures. Total losses of consumable food grains due to lack of post-harvest handling was around 12–16 million metric tons out of which nearly 1–2.4 million metric tons consumable food grains are spoiled due to inappropriate and scientific storage structures. Due to unavailability of sufficient modern storage structures, huge amount of grain has to be lost per year that is why improved structure like warehouse should be used for preserving the grain up to long duration without damage the significant amount of food grains.

Prevention of food spoilage and shelf-life of food grain without significant degradation in quality is possible with the use of appropriate advanced and improved storage structure for storing the grain. Storage of food grains enhanced the opportunity of obtaining higher price during off-season or time when those particular grains are not available in the market. Storage of grain is treated as first step to preserve the food grains because the industries were demanding good quality raw materials (like grains and seeds) for manufacturing the final consumable products. Regular inspection and quality control practices during storage are important necessity to minimize the stored grain damage and facilitate the sufficient supply to industries for further used.

KEYWORDS

- Advanced storage structures
- Agricultural produces
- Air tight storage systems
- Banda
- Bin
- Biological activities
- Bukhari
- Bulk storage
- Capacity
- Cold storage
- Consumption
- Controlled atmospheric storage
- Controlled atmospheric storage system
- Conventional storage structures
- Cover and plinth storage structure
- Crops
- Cultivation
- Dokal
- Domestic food grain storage structures
- Domestic storage structures
- Earthen pot
- Economical
- Environmental conditions
- Food
- Food Corporation of India
- Food grain
- Food grain production
- Food grain safely
- Grain
- Grain storage
- Gummi
- Hagevu
- Hapur tekka
- Hermetic storage
- Hermetic storage system
- Hermetic storage technology
- Improper storage
- Indigenous storage structures
- Indoor storage structures
- Insect-pest
- Insect-pest attack
- Insects
- Insects-pest losses
- Kanaja
- Khatties
- Kothi
- Kuthar
- Kuthla
- Losses
- Major losses
- Microbial activities
- Microbial attack
- Microbial losses
- Micro-organism
- Modern storage structures
- Modified atmosphere
- Modified domestic clay bins
- Morai
- Mud bin
- Pallets
- Pau bin
- Pit
- Post-harvest
- Post-harvest handling
- Post-harvest management
- Puri

- Pusa bin
- Qualitative deterioration
- Qualitative loss
- Quality
- Quantitative loss
- Rodents
- Safe storage
- Sanduka
- Semi underground cement bin
- Silo
- Silos
- Small storage
- Socio-economic
- Socio-economic status

- Spoilage
- Storage
- Storage management
- Storage practices
- Storage structure
- Storage system
- Stored food grain
- Thekka
- Traditional crib
- Traditional storage structures
- Traditional storage systems
- Underground storage structures
- Warehouse

REFERENCES

1. Acharya, S. S., & Agrawaal, N. L., (2009). *Agriculture marketing in India*. Oxford and IBH Publishing Company Pvt. Limited. New Delhi, India, pp. 111–116.
2. Aldred, D., & Magan, N., (2004). Prevention strategies for tricothecenes. *Toxicology Letters, 153*, 165–171.
3. Bogaard, A., Charles, M., Twiss K. C., Fairbairn A., Yalman N., & Filipovic D., (2009). Private pantries and celebrated surplus: storing and sharing food at Neolithic Catalhoyuk, Central Anatolia. *Antiquity, 83*(321), 629–668.
4. Campos, M. G., (2006). Good practices in grain storage, stored grain procedures and practices. In: Lorini, I., Bacaltchuk, B., Beckel, H., Deckers, D., Sundfeld, E., dos Santos, J. P., Biagi, J. D., Celaro, J. C., Faroni, L. R., Bortolini, L. O. F., Sartori, M. R., Elias, M. C., Guedes, R. N. C., da Fonseca, R. G., Scussel, V. M. Eds., *Proceedings of the 9th International Working Conference on Stored Product Protection*, 15-18 October, 2006, Campinas, São Paulo, Brazil. Brazilian Post-harvest Association - ABRAPOS, Passo Fundo, RS, Brazil, pp. 1295-1301. Accessed on 16 May 2016, URL: http://spiru.cgahr.ksu.edu/proj/iwcspp/iwcspp9.html
5. Channal, G., Nagnur, S., & Nanjayyanamath, C., (2004). Indigenous grain storage structures. *Leisa India, 6*(3), 10.
6. Dhaliwal, R. K., & Singh, G., (2010). Traditional food grain storage practices of Punjab. *Indian Journal of Traditional Knowledge, 9*(3), 526–530.
7. Donahaye, J., Navarro, S., Varnava, A., (1997). *Proc. Int. Conf. Controlled Atmosphere and Fumigation in Stored Products*. Printco Ltd., Nicosia, Cyprus, pp. 183–190.

8. Edwards, W., (2013). *Grain Storage Alternatives: An Economic Comparison.* Ag Decision Maker, Iowa State University of Science and Technology, Ames, Iowa. USA, pp. 130.

9. Ferizli, A. G.; Emekci, M., (2000). Carbon dioxide fumigation as a methyl bromide alternative for the dried fig industry. In: Obenauf, G. L. and Obenauf, R. Eds., *Annual International Research Conference on Methyl Bromide Alternatives and Emissions Reductions*, Orlando, Florida, pp. 81.

10. Flinn, P. W., Hagstrum, D. W., & Muir, W. E.. (1997). Effects of time of aeration, bin size, and latitude on insect populations in stored wheat: a simulation study. *Journal of Economic Entomology, 90*, 626–651.

11. Flinn, P. W., & Subramanyam, B., & Arthur, F. H., (2004). Comparison of aeration and spinosad for suppressing insects in stored wheat. *Journal of Economic Entomology, 97*, 1465–1473.

12. Groot, I. D., (2004). *Protection of Stored Grains and Pulses*, Agrodok 18, Agromisa Foundation, Wageningen, Netherlands, pp. 8–20.

13. Gupta, S. L., Singh, S. P., & Dubey, V. K., (1994). Traditional Wisdom: A Conceptual Exploration, *Interaction, 12*(3), 3–25.

14. Hall, D. W., (1970). *Handling and storage of food grains in tropical and subtropical areas.* FAO Agricultural Development, Paper 90, Rome, Italy, pp. 140.

15. Hell, K., Cardwell, K. F., & Setamou, M., (2000). The influence of storage practices on aflatoxins contamination in maize in four agro ecological zones in Benin. *Journal of Stored Products Research, 36*, 365–382.

16. Holscher, K., (2000). Integrated pest management of stored grain insects: current status and future concerns. *Proceeding of 12th Annual Integrated Crop Management Conference,* Iowa State University, Lowa City, USA, pp. 41–47.

17. Jain, D., Satapathy, K. K., & Wahlang, E. L., (2004). *Traditional Post Harvest Technology of North Eastern Hill Region.* ICAR Research Complex for NEH Region, India, pp. 134.

18. Jakob, G., Dierks-Lange, H., Heck, F. W., & Schmitt, S., (2006). The speedbox – an innovative application device for the Degesch plates. In: Lorini, I., Bacaltchuk, B., Beckel, H., Deckers, D., Sundfeld, E., dos Santos, J. P., Biagi, J. D., Celaro, J. C., Faroni, L. R. D. A., Bartolini, L, de, O. F., Sartori, M. R., Elias, M. C., Guedes, R. N. C., De-Fonseca, R. G. and Scussel, V. M. Eds., *Proceedings of the Ninth International Working Conference on Stored Product Protection,* 15–18 October, Sao Paulo, Brazil, Brazilian Postharvest Association, Campinas, Brazil, pp. 554–556. Accessed on 16 May 2016, URL: http://spiru.cgahr.ksu.edu/proj/iwcspp/iwcspp9.html

19. Kanwar, P., & Sharma, N., (2006). Traditional storage structures prevalent in Himachali homes, *Indian Journal of Traditional Knowledge, 5*(1), 98–103.

20. Kanwar, P., & Sharma, N., (2003). An insight of indigenous crop storage practices for food security in Himachal Pradesh. In: *Kanwar, S. S., Sardana, P. K. and Satyavir, K. Eds., Food and Nutritional Security, Agrotechnology and Socio-economic Aspects.* SAARM, India, pp. 230.

21. Karthikeyen, C., Veeraragavathatham, D., Karpagam, D., & Ayisha Firdouse, S., (2009). Indiginous storage structures. *Indian Journal of Traditional Knowledge, 8*(2), 225–229.

22. Kendall, H. W., & Pimentel, D., (1994). Constraints on the expansion of the global food supply. *Ambio, 23*, 198–205.
23. Kimatu, J. N., McConchie, R., Xie, X., & Nguluu, S. N., (2012). The Significant Role of Post-Harvest Management in Farm Management, Aflatoxin Mitigation and Food Security in Sub-Saharan Africa. *Greener Journal of Agricultural Sciences, 2*(6), 279–288.
24. Kiruba, S., Manohar Das, S. A., & Papadopoulou, S., (2006). Prospects of traditional seed storage strategies against insect infestation adopted by two ethnic communities of Tamil Nadu, southern peninsular India. *Bulletin of Insectology, 58*(2), 129–134.
25. Lorini, I., (1993). *Aplicação do Manejo Integrado de Pragas em Grãos Armazenados. In: Anais do Simpósio de Proteção de Grãos Armazenados*, Passo Fundo, INR EMBRAPA-CNPT, pp. 147.
26. Magan, N., & Aldred, D., (2007). Post-harvest control strategies: minimizing mycotoxins in the food chain. *International Journal of Food Microbiology, 119*(1–2), 131–139.
27. Mehrotra, S. N., Verma, N., Datta, A., & Batra, Y. K., (1987). *Building research note, Central building research institute, India, small capacity grain storage bins for rural areas*. Central Building Research Institute. Roorkee, India, pp. 120.
28. Minjinyawa, Y., (2010). *Food and Crop Storage Technology, Farm structures*. 2nd Edition. Ibadan University Press, Ibadan, Nigeria, pp. 110.
29. Mortimore, S., & Wallace, C., (1994). *HACCP: Practical Focus*. 1st Ed., Chapman & Hal. UK, pp. 289.
30. Mtiraichelvan, R., & Raman, K. V., (1990). Indigenous knowledge of farmers: Its use in extension strategies for rain fed agriculture. In: *Singh, S. P. and Prasad, C. Eds., Technologies for Minimizing Risk in Rainfed Agriculture*. ISEE and IFAD and ICAR, New Delhi, India, pp. 110.
31. Naik, S. N., & Kaushik, G., (2011). *Grain Storage in India: An Overview*. Centre for Rural Development & Technology, IIT Delhi. India, pp. 119.
32. Navarro, S.; Noyes, R. *The Mechanics and Physics of Modern Grain Aeration Management* (Navarro, S. and Noyes, R. Eds.). CRC Press, Boca Raton, FL, pp. 627.
33. Navarro, S., Donahaye, E., Kashanchi, Y., Pisarev, V., & Bulbul, O., (1984). Airtight storage of wheat in a PVC covered bunker. In: *Ripp, B. E. et al., Eds. Controlled Atmosphere and Fumigation in Grain Storages*. Amsterdam, Elsevier, pp. 591–604.
34. Page, A. B. P., Lubatti, O. F., (1962). Fumigation of insects. *Annual Review of Entomology, 8*, 239–264.
35. Phillips, T. W., Throne, J. E., (2010). Biorational approaches to managing stored-product insects. *Annual Review of Entomology, 54*, 375–397.
36. Proctor, D. L., (1994). Grain storage techniques - Evolution and trends in developing countries. FAO Agricultural Service Bulletin 109, pp. 233.
37. Ravi, K., Jahaginder, S. R., Siddaramiah, K., (2002). Traditional methods in the management of plant disease. *Indian Farmer Digest, 35*, 29–32.
38. Reed, C. R., Hagstrum, D. W., & Flinn, P. W., (2003). Allen, R. F. Wheat in bins and discharge spouts, and grain residues on floors of empty bins in concrete grain elevators as habitats for stored-grain beetles and their natural enemies. *Journal of Economic Entomology, 96*, 996–1004.
39. Rezende, A. C., (2002). Good storage practices: hazard analysis and critical control points. In: *Campinas, S. P. Ed., Grain Storage*. Bio Genezis, pp. 1000.

40. Sabio, G. C., Dator, J. V., Orge, R. F., Julian, D. D. T., Alvindia, D. G., & Miranda, G. C.; et al., (2006). Preservation of Mestizo 1 (PSB Rc72H) seeds using hermetic and low temperature storage technologies. In: *Lorini, I., Bacaltchuk, B., Beckel, H., Deckers, D. (Eds.), Proceedings of the 9th International Working Conference on Stored Product Protection*, 15-18 October, 2006, Campinas, São Paulo, Brazil. Brazilian Post-harvest Association - ABRAPOS, Passo Fundo, RS, Brazil, pp. 946–955. Accessed on 16 May 2016, URL: http://spiru.cgahr.ksu.edu/proj/iwcspp/iwcspp9.html

41. Sahay, M. K., & Singh, K. K., (2009). *Unit Operations of Agricultural Processing*. Vikas Publishing House Private Limited, 2nd Ed., Noida, India, pp. 169–173.

42. Said, P. P., & Pradhan, R. C. (2014). Food Grain Storage Practices - A Review. *Journal of Grain Processing and Storage*, *1*(1), 01–05.

43. Sashidhar, R. B., Ramakrishna, Y., & Bhat, R. V., (1992). Moulds and mycotoxins in sorghum stored in traditional containers in India. *Journal of Stored Products Research*, *28*(4), 256–259.

44. Schmutterer, H., (1990). Properties and potential of natural pesticides from the neem tree (*Azadirachta indica*). *Annual Review of Entomology*, *35*, 71–297.

45. Shah, D., (2013). *Assessment of Pre and Post-Harvest Losses in Tur and Soybean Crops in Maharashtra, AERC Report*. Agro-Economic Research Centre, Gokhale Institute of Politics and Economics, Pune, India, pp. 119.

46. Singh, P. K., (2010). A decentralized and holistic approach for grain management in India. *Current Science*, *99*(9), 1179–1180.

47. Singh, R. K. P., & Satapathy, K. K., (2003). Zero Energy cool chamber: a low cost storage structure. *Journal of the North Eastern Council*, *23*(3), 27–30.

48. Sinha, A. K..; & Sinha, K. K., (1990). Insect pests, *Aspergillus flavus* and aflatoxin contamination in stored wheat: a survey at north Bihar (India). *Journal of Stored Products Research*, *26*(4), 223–236.

49. Srivastava, S., Gupta, K. C., & Agrawal, A., (1988). Effect of plant product on *Callosobruchus chinensis* L. infection on red gram. *Seed Research*, *16*(1), 98–101.

50. Tripkovic, B., Containers and grains: food storage and symbolism in the Central Balkans (Vina period), Prehistoric document XXXVIII, 2011. Accessed on 16 May 2016, URL: http://arheologija.ff.uni-lj.si/documenta/pdf38/38_13.pdf

51. Varnava, A., (2002). Hermetic storage of grain in Cyprus. In: *Batchelor, T. A., Bolivar, J. M. (Eds.), Proceeding of International Conference on Alternatives to Methyl Bromide*, March 5-8, 2002, Sevilla, Spain, pp. 162–168.

52. Villers, P., De Bruin, T., & Navarro, S., (2006). Development and Applications of the Hermetic Storage Technology. In: Lorini, I., Bacaltchuk, B., Beckel, H., Deckers, D., Sundfeld, E., dos Santos, J. P., Biagi, J. D., Celaro, J. C., Faroni, L. R., Bortolini, L. O. F., Sartori, M. R., Elias, M. C., Guedes, R. N. C., da Fonseca, R. G., Scussel, V. M. (Eds.), Proceedings of the 9th International Working Conference on Stored Product Protection, 15-18 October, Campinas, São Paulo, Brazil. Brazilian Post-harvest Association - ABRAPOS, Passo Fundo, RS, Brazil. pp.719–729. Accessed on 16 May 2016, URL: http://spiru.cgahr.ksu.edu/proj/iwcspp/iwcspp9.html

53. Villers, P., Navarro, S., & De Bruin, T., (2008). Development of Hermetic Storage Technology in Sealed Flexible Storage Structures Controlled Atmosphere and

Fumigation (CAF). Conference Paper, Grain Pro, Inc. Organic Storage Systems, Accessed on 16 May 2016, URL: http://grainpro.com/gpi/images/PDF/Commodity/CAF_Presentation_Development_of_Hermetic_Storage_Technology_PU2015PV0708.pdf

54. Weaver, D. K., & Petroff, A. R., (2004). Pest management for grain storage and fumigation, Montana Department of Agriculture. Accessed on 16 May 2016, URL: http://pesticides.montana.edu/reference/documents/FumSeed.pdf

55. Weinberg, Z. G., Yan, Y., Chen. Y., Finkelman, S., Ashbell, G., & Navarro, S., The effect of moisture level on highmoisture maize (*Zea mays* L.) under hermetic storage conditions-in vitro studies. *Journal of Stored Products Research, 44,* 136–144.

56. White, N. D. G., & Jayas, D. S., (2003). Quality changes in grain under controlled atmosphere storage. In: *Navarro, S. and Donahaye, E. (Eds.), Proceedings of the International Conference on Controlled Atmosphere and Fumigation in Grain Storages,* Caspit Press Limited, Jerusalem, Israel, pp. 205–214.

GLOSSARY OF TECHNICAL TERMS

Active metabolism is a phenomenon of the body to support metabolism in which multivitamin received from specially formulated dietary supplement with ingredients (such as vitamins, dietary minerals, and other nutritional elements) are used by body to converts into energy.

Adulteration is a process to make any material impure by missing any impure or harmful ingredients in it. In the term of food processing and agriculture, it is defined as a procedure of mixing some low quality or impure materials which hampers the quality and nutritional value of original food materials

Advanced storage structure refers to scientific and advance technologies based different storage structures made with various materials like bricks, concrete, fiber, polythene and metal etc., for providing safe storage by preventing the attack of insect-pest, rodents and birds etc.

Alkaloids refers to different organic compounds contain at least one atom of nitrogen and frequently find in plants.

Ampicillin is a β-lactam antibiotic used against different type bacterial infections and diseases (such as endocarditis, meningitis, respiratory tract infections, salmonellosis and urinary tract infections) to prevent and control them.

Anthocyanidins refer to pigment compounds found in plants which are free from sugar.

Anthocyanins are members of the flavonoid family of polyphenol phytochemicals found in various plant foods. They are naturally occurring water-soluble vacuolar pigments compounds in plants in the form of glycosides in which a sugar is paired by an anthocyanidin molecule.

Anti-aging is defined as a technique which slows down or stops the effect of aging and prevents becoming older.

Antibacterial is a chemical substance either natural or synthetic that kills, reduces, counteracts or eliminates bacteria from growing is antibacterial agent.

Antidiabetics are used to control the levels of blood sugar in diabetes.

Antifungals are chemicals used to treat fungal infection of the skin, hair and nails.

Antioxidant is a material or substance prevents the deterioration of any food material by slow down the oxidization process. It is generally available in natural or organic nutritious food materials and beneficial for retarding health decay.

Anti-plaque agents have to have ability that is anti-bacterial effect on the growth of plaque bacteria.

Apiary is defined as a small area or place where numbers of beehives has been kept for obtaining honey in collective amount.

Apiculture is the science and art of multiplication and care of bees. It is also known as Beekeeping.

Apitoxin is a colorless venom extracted from honey bees used for pharmaceutical purpose to cure many diseases like rheumatic diseases and arthritis etc.

Aroma is a specific or distinctive pleasant or savory smell noticed from each and every food products or any other materials.

Asparagines refers to an α-amino acid (non-essential), generally obtained from plants and other sources and white in color. It is used in biosynthesis of protein and sometime also used as culture medium for some specific bacteria. It is soluble in water also.

Azeotrope refers to a mixture of two or more liquids when boiled, the vapors have same ratio of constituent liquids as that of unboiled mixture.

Bamboo splits refer to vertical segments or pieces of a wholesome bamboo called bamboo splits. It's generally used for constriction of traditional storage structures and sometime also used for making rural houses and cattle shed etc.

Beehives are a container in which honey bees has been kept and honey collection taken place. Generally it is rectangular, square and dome shape.

Beekeeper refer to persons who are involved in Beekeeping or Apiculture for collecting the honey products like honey, wax, pollen, brood, venom and royal jelly etc.

Beeswaxes are produced by bees, contains high amount of fat and used for preparation of candles, polishes and other materials.

Beverage is any consumable drink except water known as beverages. Generally it is prepared for commercial purpose and supplied in market as energy, refreshing, medicinal and nutrients etc., drinks e.g. Coffee, Tea, Health drinks and Chocolate drinks etc.

Bin is a circular, rectangular or square shape structure with vacant enclosed space for storing the farm products like grains etc.

Bioactive compounds are substances with biological activities and affecting the cells, tissues and other organs of living being.

Biochemical is a chemical process and reaction occurring in living beings. Photosynthesis, respiration or oxidation etc., all are consider as biochemical process either in plant or in human beings.

Black tea is a type of tea which is prepared by fresh and fully fermented and then dried tea leaves. This is more widely consumed in western countries. Generally it is milk less tea.

Blanching is a process, formally known as mild heat treatment, in which fruit or vegetables are frequently employed to heat (up to the desired temperature) for a short period of time into the water prior to essential step like canning, drying or freezing. It is done primarily for the purpose of inactivating enzymes and also preserves the colour of fruit and vegetables.

Bleeding gums refer to the formation of plaque (such as debris and bacteria associate/sticks with teeth) at the gum line is main reason of bleeding gums. They are responsible and lead to gingivitis condition.

Bunker is defined as a underground structure with strong and sealed wall used for storage of food grain in bulk amount up to long duration.

Caffeine is a chemical substance (alknoids), most commonly available in beverage like Coffee, tea and chocolates and act as stimulant. Chemical formula is $C_8H_{10}N_4O_2$. *Caffeine* makes people more alert and less drowsy etc., because its stimulates the central nervous system.

Calcified tissues are found the teeth. In dental caries, it is formed as localize dissolution and damaged by cariogenic bacteria.

Cancer is a group of diseases begins with uncontrolled growth of cells in human body and this cells are able to transfer to distant site. It is have different ypes like blood cancer, skin cancer, lung cancer and breast cancer etc.

Canning is a thermal processing (also known as in-container sterilization) is used to preserve fruits, vegetables, food and their products (cooked or uncooked food) by sealing them in the airtight container prior to sterilization process with high temperature. The objective of canning is to eliminate microbes associated with spoilage and public health issue.

Caramelization is a procedure in which color of food items are changed into light or deep brown due to heating of sugar present in it. During heating, volatile compounds released and causing a different flavor and essence in food. It is also known as browning of sugar.

Cariogenic bacteria are teeth decaying bacteria e.g cavities, inflammation of tissue around the teeth or break down of teeth etc.

Cariogenic refers to promoting or prucing the causes for tooth decay by facilitating the acid forming bacteria on teeth.

Carotenoid is an organic pigment found in the chloroplasts and chromoplasts of fruits and vegetables, generally responsible for yellow, bright red, and orange color. In addition, it have to have key role in plant health i.e. very important.

Catechin is a crystalline, astringent and water soluble flavonoid compound (chemical formula-$C_{15}H_{14}O_6$) or its derivatives having antioxidant properties and used in dyeing and tanning.

Chaptalization is the process of adding sugars to fermenting grape juice so as to increase the alcohol content of final product. The process so called after the name of its developer Jean-Antoine-Claude Chaptal.

Chewable tablets are important alternative and convenient to conventional tablets in dental diseases. Such tablets have great and potential merits because no need of water. It can be given to the patient at any place in any time. Chewable tablets provides additional advantages for patients such as ensuring better compliance, improving the experience and overcome swallowing difficulties.

Chromatogram refers to the pattern observed on an adsorbent medium (column, gel layer or paper) by the chromatographic separation of individual components of a mixture.

Cognitive disorder is a disorder that affects the cognitive functions (like hearing, reading, remembering and thinking etc.) and later extended in terms of dementia and amnesia etc.

Cold storage is an artificially made low temperature room or structure used for keeping the perishable agricultural produces with minimized deterioration up to a certain time period

Consolidated bioprocessing refers to a process in lignocellulosic ethanol production where all four biological processes viz. production of saccharolytic enzymes, saccharification and fermentation of both hexose and pentose sugars are carried out in single reactor.

Contamination is a process to making any material poisonous or polluted. It can be taken place due to adulteration of any substance or by mix up any low quality ingredients.

Conveyors are defined as a mechanical device in which an endless belt made with rubber, metal or any other material attached with two or more pulleys used for carrying the materials (e.g. paddy and husk etc.) from place to an-other.

Decaffeinated coffee is a process of removal of caffeine from the products which have it. Generally for removal of caffeine, coffee bins are soaked in compressed carbon dioxide which absorb the extracted caffeine from coffee.

Decaffeinated tea is a process of removal of caffeine from the beverages or product which has it. Generally for removal of caffeine (alkaloid compound, Chemical formula: $C_8H_{10}N_4O_2$ which stimulate the nervous system), from tea leaves known as decaffeination.

Demineralization is a condition on the tooth surface which is caused by bacteria.

Dental caries refers to an indigenous infectious microbial disease caused by cariogenic bacteria that lead to the localized dissolution and damage of the calcified tissues of the teeth. Dental caries is also known by their different name as tooth caries, cavities, or decay.

Dental plaque refers to the sticky substance form inside the mouth with the complex combination of food debris, acid, bacteria, and saliva

Desiccant is a substance that absorbs moisture from air and maintains dryness in its immediate environment.

Desorption is a process to release an absorbed or adsorbed substance through pores or interstices.

Diabetes is a metabolic disease associated with high levels of blood sugar (blood glucose) that cannot be used properly by body. It may cause by inadequate production of insulin, known as type I while type II diabetes characterized by insulin resistance.

DNA injury is a defect or deviation in DNA due to some physical and chemical agents and takes place in natural or in-natural both ways. Some environmental factors such as ultraviolet light and other radiations can cause *DNA damage*

Domestic food grain storage structure is defined as small house hold size of traditional or modern storage structures made with various materials like either locally available materials (e.g. mud, bamboo, cow dung and paddy straw etc.) or long durable and costly materials (e.g. bricks, concrete, fibre, polythene and metal etc.), used safe storage by preventing the attack of insect-pest, rodents and birds etc.

Downstream processing refers to series of steps undertaken for the recovery and refining of alcohol from fermentation broth.

Drying is a thermo-physical action having its dynamic principles governed by heat and mass transfer laws, following the mechanism of penetration of heat within the product and releasing of moisture through evaporation in a form of unsaturated gas phase.

Enter-Doudoroff pathway refers to an alternative carbohydrate catabolic pathway, occurs only in prokaryotes using 6-phosphogluconate dehydratase and 2-keto-3-deoxyphosphogluconate aldolase with release of only 1 ATP, 1 NADH and 1 NADPH compared to Embden- Meyeroff-Parnas (EMP) pathway that occurs in both prokaryotes and eukaryotes using hexokinase, phosphofructokinase and glyceraldehyde-3-phosphate dehydrogenase enzymes with the release of 2 ATP and 2 NADH.

Entrepreneurs is a group of business men who are going to start any firm or industries by taking risk of wealth and ideas with aim to become profit making and successful businessman known as entrepreneurs

Enzyme is any individual or group of chemical substances generated from living cells or organisms and responsible for several chemical reactions or biochemical changes without change its own characteristics.

Essential oils are volatile oils, which are present in the medicinal plants.

Extraction time is the required time to extract entirely essential oils from plants.

Fermentation is a process in which material or substance breakdowns due to activity of micro-organism like yeast, bacteria or fungi etc.

Filtering gauze is a device or instrument made with woven clothes, fiber or plastic etc., used for removing the foreign particles or other kind of impurities from the main liquid products.

Flavonoids are low molecular compounds mainly present in plants and are largest class of phenol usually in the form of glycosides. They contains two aromatic rings; ring A and B that is attached with a heterocyclic ring C. On the basis of structural variation in substitution pattern to C ring, flavonoids can be subdivided into major two subclasses: flavones and flavanols. Examples of flavonoids areanthocyanidins, catechins, kaempferoland quercetin. They are found in fruits (apple, berries, citrus etc.), vegetables (carrots, broccoli, eggplant, peppers, squash, tomatoes etc.), plantation (tea and coffee) and other crops.

Flavor is a specific quality or property of materials which indicates the particular taste. Some time, other ingredients are also mixed for getting different taste.

Fluidized bed drying includes spouted beds, which have the potential to reduce the drying time compared to tray drying and if the feature of vibration is added to it, its efficiency gets increased.

Food debris is caused of dental plaque in addition to the bacteria inside the mouth.

Freeze dryer *is a* device or instrument used to remove the moisture from materials. This machine *freezes* the material through sublimation. It is used to preserve perishable products and extion of shelf life of materials.

Fumigation is a process in which pest or insect controlling chemicals are applied in form of smokes or fumes. Generally used in warehouses for disinfecting the insects and pests attack

Fungus refer to various type of unicellular, multi-cellular or mass of cytoplasm organism causing decay of living elements and organic materials. It is a type of Micro-organism.

GC-MS is a new technique used for finding chemical composition of essential oils.

Genus is defined as group of animal or plants of similar characteristics. It is a hierarchy used in the biological classification of living and fossil organisms in biology. In the hierarchy of biological classification, *genus* comes above species and below family.

Gingivitis refers to a bacterial infection which is usually responsible for an inflammation of the gums.

Glucose refers to monosachrides (simple sugar), frequently available in several carbohydrates as a part of it. Different agricultural and forest produces like crops, fruits, vegetables and dry fruits etc., also known as good source of glucose which provides energy to the living organism

Glycolysis is a metabolic process of breakdown of glucose into pyruvic acid or lactic acid with the release of energy in form of ATP.

Green tea is defined as a beverage prepared with unfermented and dried tea leaves by soaking into water. For preparing it used tea leaves should not gone through withering, oxidation and shaping. It is originated China and then spread worldwide and now days treated as one of the most popular drink.

HACCP refers to "Hazard analysis and critical control point" and is defined as a system of includes different safety and hygiene standards for controlling the qualitative degradation of food materials from receiving to final form of product.

Herbal gels refer to semi rigid systems in which movement of the dispersing medium is restricted by an interlacing three dimensional network of particles from different herbs or solvated macromolecules of the dispersed phase.

Herbal lollipop refers to a sugar-free product made up of licorice extracts using a standard sugar-free candy formula suggested by Dr. John's Candy.

Herbal mouthwashes refer to concentrated aqueous solutions of anti-microbial preparations of plant extract, routinely used in the oral cavity after dilution to counter infections, for cleansing and anti-sepsis as well as refreshing the oral cavity. The health benefits of herbal mouthwashes include, relieving symptoms of gingivitis, canker sores, inflamed gums, sore mouth, inflamed or ulcerated throat, mouth infections, bleeding gums and teeth sensitivity.

Herbal tooth powder refers to a tooth-cleaning agent that is almost entirely made from all-natural ingredients. Its purpose was to refresh breath, help heal gums, rid teeth of bacteria and reduce the amount of inflammation in the mouth.

Herbal toothpaste refers to a formulation of well constituted herbs that ensure anti-bacterial and gum tightening properties and provide absolute dental care.

Hermetic storage is defined as a low oxygen and high carbon dioxide modified concept of storage in which food grain kept within a completely tight sealed atmosphere. Technology is used for storing the food material with good shelf life up to the long duration.

Honey extractor is a device contains rectangular or square box with moving fins or blade, made with metal, fiber or some time wood used for extracting the honey from the bee combs without hampering the combs.

Honeybees are carrier of pollination causing insect having wings and sting, leaving in hives in a group. Honey bees are producing honey, pollen, waxes and royal jelly etc.

Honeycomb refers to a home of honey bees made with wax. In this wax structure, bees are storing honey and laying eggs.

Hydro-diffusion is a diffusion of hot water and essential oils via aromatic plant membranes,

Hydro-distillation is a common traditional extraction method.

Hydrolysis is a chemical reaction in which water react with a substance or compounds and form another element or compound.

Insomnia is a health disorder in which peoples are not getting sleep up to certain period of time. It can be create another health problem also if not cured within appropriate time period.

Inspection door refers a rectangular, circular or square shape door made with wood, glass or metal can slid or open. Generally used in godown or warehouses for observing the condition of food grains.

Labelling machine is an instrument used to put the information tag, label or seal on the material packets, containers or on products which contains all the necessary information about that product or material.

Livelihood is defined as a way to earning of necessity of bio lives includes food, house and clothes etc. for survival

Lysine refers to high protein content amino acid very essential in diet for proper growth of bones.

Macronutrients refer to a class of chemical compounds, also known as macronutrient such as carbohydrates, fats, protein etc. is used as a major portion for consumption to contribute energy to our body. Macronutrients include, and alcohol.

Mannose refers to a hexose sugar with a chemical formula:$C_6H_{12}O$and frequently available in many polysaccharides. *It* is have vital role in human metabolism (e.g. glycosylation of certain proteins) and use in preventing urinary tract infections (UTIs).

Melanoidins is a brown polymer (high molecular weight), produced during reaction of amino acids and sugar at high temperature (reaction is known as Millard reaction).

Micronutrients are chemicalsubstances (such as vitamins and mineral) that isessential for proper growth, development and metabolism of a living organism. They are required in minute amounts utilized by the body.

Micro-organisms refer to very small living organism with single and multi-cellular cells able to observe with micro-scope. Fungi, Virus, algae and bacteria etc., considered as micro-organisms

Microwave is an apparatus used to heat the mixture of water and plant via electromagnetic waves.

Moisture content is a ratio of amount of water present and amount of solid present in the any material or substance. Generally it is evaluated on dry and wet basis in percentage.

Molasses refer to viscous light to dark brown colored syrup obtained as a byproduct during refining of sugarcane or sugar beet into sugar.

Mycotoxin refers to a poisonous chemical substance released from some certain variety of fungi. Generally it is occurred in stored food grains in unhygienic way.

Nectar is a sweetened liquid extracted from flowers or any plants by honey bees and enhance the pollination and finally converted into honey.

Ohmic heating is a novel technique for pasteurizing milk via passing electric current through food.

Ohmic-hydro-distillation is an advanced method using ohmic heating process and could be considered as a novel method for the extraction of essential oils.

Oil yield represents the produced essential oils from the weight of plant.

Organoleptic analysis refers to the analysis of alcohol using sense organs like taste, sight, aroma, texture astringency etc.

Oxidant is a reactant that oxidizes another substance.

Oxidation is process in which oxygen reacts with elements or compounds and altered the appearance of substances or elements.

Packaging machine is an assembly of instruments used for complete packaging of food or any other materials includes filling, sealing, wrapping and labeling etc.

Pallet refers to flat rectangular or square wooden structure used to keep the food grain bags over on it.

Pasteurization is a process, named after scientist Louis Pasteur, where heat is applied to destroy pathogens in foods. It is a form of partial sterilization of a substance and especially for a liquid sample.

Pectin is a water soluble colloidal carbohydrate of higher molecular weight. It is frequently available in apple, plum, grape and pomegranate etc.

Phenolic compoundsrefer to aromatic compounds and available in many essential oils. It has very important medicinal properties also like good anti-bacterial and anti-septic properties etc.

Phytochemicals are the bioactive or chemical compounds which are naturally present in plants and in which some are consumed for their color, some for flavor and taste; e.g. deep purple color of blueberries, or the smell of garlic etc.

Pollination is a complete procedure of carrying the male sex cells of plants or any other vegetative crop to female reproductive organs for fertilization.

Post-harvest losses refer to accumulated losses of agricultural produces from harvesting till consumption. Food grain losses during harvesting, handling, processing, storage, Transportation and packaging etc. considered under Post-harvest losses.

Preservation of foods is a need based approach because the perishable food products get spoiled soon. They are designed to slow the growth of microorganisms and is affected by food factors both intrinsic and extrinsic factors like pH, oxidation-reduction potentials, water activity, and natural antimicrobials. Preservation of food and retention of nutrition is through all the stages of processing. The various nutrients like vitamins, proteins, amino acids, minerals etc. are sensitive to methods used in preservation.

Processing technology is procedure or method of preserving the materials with minimum losses and high shelf life. It includes all the techniques and steps from after harvesting till consumption e.g. handling, transporting, cleaning, drying, packaging and other value addition processes.

Respiration is a biological process of breathing in which oxygen has been received and carbon dioxide released. It is necessary for metabolism of living organisms.

Roasting is a method of cooking in which food materials are cooked without any liquid (e.g. oil, water or any other fluids). It is also called dry cooking.

Semi underground is rectangular, circular, semi-circular or any other suitable shape structure which half portion should be above and an-other half within the earth, generally use for storing the food grains etc., safely up long duration.

Sensory characteristic is a scientific and experimental system for analysis of food quality with help of five senses like sight, smell, taste, touch and hearing. Generally this experimental process used for consumer analysis.

Silo is a cylindrical, rectangular or square shape vertical structure with varies in height used for storage of food grains, pulses and other agricultural crops safely up to a certain period of time.

Smoker is a device used to produce smoke in beehives to keep calm to bees during extraction of honey or any other products from it.

Solar wax melter is a device made with wooden box and glass receives heat from sun radiation which is used to melt down the bee wax by keeping inside of this device.

Specific gravity is the ratio of density of alcohol to density of water at the same temperature and pressure.

Steam distillation is a widespread method for isolating essential oils commercially.

Sterilization is a process with heat under the high pressure is done to retort microorganisms that could be a health hazard or cause the food to spoil and retain optimum quality of the preserved fruits and vegetables when at their peak of freshness. In addition, sterilization also inactivates the enzymatic activity that may the reason of undesirable changes in the color, flavor and texture.

Strained honey is raw and pure honey obtained after removal of mixing ingredients like waxes etc.

Synthetic drugs are associated with some issues for example; drug resistance, high treatment cost and side effects etc.

System performance coefficient is defined as the ratio of taken heat to the given energy.

Testing gauze is an instrument or device used for checking the different parameters of honey collected from bees.

Thermo-tolerant refers to organisms, which generally grow at moderate temperature but can survive, not thrive, high temperature.

Toxins: Metabolites are produced naturally by variety of pathogens that are capable of causing disease on contact with or absorption by body tissues *via* animal or human feed or food stuff.

Traditional storage structure refers to small structures made with locally available materials like bamboo, paddy straw, cow-dung, mud and stones etc. for storing grain safely up to certain period of time.

Vapor density is the density of gas or vapors in relation to the hydrogen at same temperature and pressure. Substances lighter than air have vapor density < 1 and heavier than air have vapor density > 1.

Volatile refers to one that evaporates readily at normal temperature and pressure.

Water distillation is used to extract of essential oils from raw or dried plants by diffusion mechanism.

Whey refers to thin and watery liquid left after curdling of milk in cheese and yogurt manufacturing.

INDEX

Milton Keynes UK
Ingram Content Group UK Ltd.
UKHW022045141024
449569UK00022B/818

9 781774 636411